Positive Evolutionary Psychology

Positive Evolutionary Psychology

Darwin's Guide to Living a Richer Life

GLENN GEHER, PHD
NICOLE WEDBERG, MA
State University of New York at New Paltz

OXFORD
UNIVERSITY PRESS

OXFORD
UNIVERSITY PRESS

Oxford University Press is a department of the University of Oxford. It furthers
the University's objective of excellence in research, scholarship, and education
by publishing worldwide. Oxford is a registered trade mark of Oxford University
Press in the UK and certain other countries.

Published in the United States of America by Oxford University Press
198 Madison Avenue, New York, NY 10016, United States of America.

First issued as an Oxford University Press paperback, 2022

CIP data is on file at the Library of Congress
ISBN 978-0-19-064712-4 (hardback)
ISBN 978-0-19-765679-2 (paperback)

3 5 7 9 8 6 4 2

Paperback printed by Marquis, Canada

Contents

IV. IMPLICATIONS AND THE FUTURE OF POSITIVE EVOLUTIONARY PSYCHOLOGY

Foreword

After reading *The Origin of Species*, the prominent Cambridge geologist Adam Sedgwick wrote the following letter to Charles Darwin, his esteemed former student and field assistant:

> If I did not think you a good tempered & truth loving man I should not tell you that . . . I have read your book with more pain than pleasure. Parts of it I admired greatly; parts I laughed at till my sides were almost sore; other parts I read with absolute sorrow; because I think them utterly false & grievously mischievous. . . . There is a moral or metaphysical part of nature as well as a physical. A [person] who denies this is deep in the mire of folly. Tis the crown & glory of organic science that it *does* thro' *final cause*, link material to moral . . . you have ignored this link. . . . Were it possible (which thank God it is not) to break it, humanity in my mind, would suffer a damage that might brutalize it—& sink the human race into a lower grade of degradation than any into which it has fallen since its written records tell us of its history. (Darwin Correspondence Project, November 24, 1859)

Sedgwick was reacting strongly to his personal beliefs about the Divine creation of life, by "a power I cannot imitate or comprehend—but in which I believe, by a legitimate conclusion of sound reason drawn from the laws of harmonies of nature" (Himmelfarb, 1959/1996, p. 352). As we now know, Darwin's theory of natural selection offers the most plausible mechanism of the development of life on this planet, and the moral realm is just as much a part of nature as the physical realm. The field of evolutionary psychology has taken Darwin's radical insights and has shed important light on many aspects of human behavior—the good, the bad, and the downright ugly.

Nevertheless, the predominant focus in mainstream evolutionary psychology, up to this point, has been on the less-than-savory sides of our nature. A look at evolutionary psychology's premier textbook reveals a table of contents in which everything is phrased in terms of "problems" or "challenges" that humans faced throughout our evolutionary history: problems of survival, challenges of sex and mating, challenges of parenting and kinship, and problems of group living (Buss, 2014). The one topic that seems to point to something particularly uplifting about our species—"cooperative alliances"—is juxtaposed with topics such as combating the hostile forces of nature, aggression, and warfare; the reason why mothers provide more parental care than fathers; male and female differences in short-term and long-term sexual strategies; and individual differences in rape proclivity.

Let's contrast this with one of the premier textbooks in the field of positive psychology, which includes the following topics: the role of culture in developing strengths and living well, positive emotions, self-efficacy, optimism, hope, wisdom, courage, mindfulness, flow, spirituality, empathy, altruism, gratitude, forgiveness, and love (Lopez, Pedrotti, & Snyder, 2014). If I were an observer from another planet and read these two texts, I'd think we were talking about two fundamentally different species!

But of course we aren't talking about different species. Humankind has the capacity for *all of it*; we can be just as merciless and savage as we can be loving and creative. If evolution sculpted our brains to solve problems and overcome reoccurring challenges faced throughout human history, it most certainly also sculpted our brains to be resilient and to have extraordinary strengths of character.

Indeed, this is the focus of positive psychology: the positive character strengths, positive experiences, positive relationships, and positive institutions that enable us to flourish (e.g., Dunn, 2017; Peterson, 2006; Seligman, 2012). Recent research and books by positive psychologists have been highlighting the evolutionary forces operating on these lighter sides of human nature. For instance, Dacher Keltner, founder of the Greater Good Science Center at UC Berkeley, argued that humans are "born to be good," and that we are wired for kindness, love, compassion, and awe (Keltner, 2009). Likewise, George Vaillant argued that humans underwent a "spiritual evolution" throughout the course of evolution, and we are wired for faith, love, hope, joy, forgiveness, compassion, awe, and mystical illumination (Vaillant, 2009).

The current book follows in this tradition by spearheading an entirely new field, *positive evolutionary psychology*, focused on the evolutionary foundations of the good life. I found this book to be an incredibly refreshing, integrative, forward-looking treatise on what humans *could be*, without ignoring *what we are*. For too long, the field of evolutionary psychology has focused on the muck, ignoring or giving less airplay to the uniqueness of humanity and our immense capacities for growth, love, peace, imagination, creativity, and reason. By the same token, for too long the field of positive psychology has focused on the positive qualities of our humanity but has swept real human struggles under the rug, including the very real problems of survival and reproduction. Such an integration is long overdue, and Glenn Geher and Nicole Wedberg are to be commended for what they have produced. I know that I personally consider it a great honor to be able to write the foreword to such a forward-thinking book.

The central question the book aims to answer is this: *Can evolutionary psychology provide guidance for living a richer and healthier life?* A very admirable question! The authors aren't attempting only to describe findings, but also to *prescribe*. This is clearly in the spirit of wanting to help increase quality of life, a very worthy goal, and one that is shared with positive psychology. The authors put forward one potential way that evolutionary psychology can inform the cultivation of the good life: understanding ways in which our modern environments are mismatched from the conditions under which our bodies and minds originally evolved. As they rightly point out, humans in Westernized societies experience many important instances of evolutionary mismatch, such as seen in education, politics, and the overuse of technology instead of real human connection. The authors show how such an understanding can help us intentionally design institutions and practices that bring out the best, not worst, in humans.

There is certainly promise to this approach. As clinical psychologist Stephen Ilardi has put it, "We were never designed for the sedentary, indoor, sleep-deprived, socially-isolated, fast-food-laden, frenetic pace of modern life" (http://tlc.ku.edu). His treatment for depression, Therapeutic Lifestyle Change (TLC), is an effective treatment for depression, with at least 50% of patients showing symptom reduction. Part of his treatment involves living the lifestyle our ancestors lived, with a balanced diet, exercise, sunlight, sleep, and social support (Ilardi, 2010).

While I agree with the authors of this book that an increased understanding of the powerful forces of natural selection can provide some hints on how one can live a life of optimal well-being, I'd like to offer some suggestions that I hope can help move the field even further forward. Geher and Wedberg cover a wide gamut of topics in this book, ranging from cooperation, to kindness, to religion, to happiness, to gratitude, to resilience. These are indeed topics that are being investigated in positive psychology. However, one topic is notably absent from this list: *meaning*.

One finding coming from positive psychology is the need to distinguish between a "happy life" and a "meaningful life" (Baumeister et al., 2013).

Certainly the two overlap, but there are important differences. Satisfying one's needs and wants tends to increase happiness, but tends to be largely irrelevant for meaningfulness. Happiness also tends to be present-oriented, whereas meaningfulness tends to involve integrating past, present, and future. In fact, thinking about future and past is associated with high meaningfulness but *low* happiness. Happiness is also linked to being a taker rather than a giver, whereas meaningfulness is the opposite. Higher meaningfulness can even lead to higher levels of worry, stress, and anxiety while resulting in lower levels of happiness. Yet, many humans certainly include meaning as a core part of the good life.

Even within the happy life, there is reason to be skeptical about blindly trusting our positive emotions, even if they are "evolutionarily adaptive." Evolutionarily adaptive doesn't necessarily mean healthy, happy, or socially desirable. Things can be conducive to reproductive fitness without necessarily contributing to personal well-being. Our genes don't care about our happiness or mental health; their primary reason for existing is to propagate themselves into the next generation. One implication of this idea is that positive emotions shouldn't unquestioningly be our guide to the good life. Just because we discover that some set of behaviors made our ancestors feel good doesn't mean that's how you, as a *whole organism*, would want to live your own life. There are many behaviors that make us feel good in the moment but thwart our longer-term goals, or even hurt others in the moment. We have many competing goals and "selves," as evolutionary psychologists Robert Kurzban (2012) and Douglas Kenrick and Vladas Griskevicius (2013) have astutely pointed out.

The key to living a good life, in my view, isn't just following whichever impulse makes us feel good, or even made our distant *ancestors* feel good, but harnessing what's best within us and regulating our impulses so that we can get closer and closer to those higher-level goals. Consider Kenrick, Griskevicius, Neuberg, and Schaller's (2010) hierarchy of needs, in which they placed parenting at the top of their hierarchy, above "mate retention," "mate acquisition," "status/esteem," "affiliation," "self-protection," and "immediate physiological needs." This may be a fair ordering of priorities for our genes, but this is hardly a fair ordering of priorities for whole organisms. Any model of human needs that is exclusively "built upon ancient foundations" is sure to be largely unsatisfying to human beings. When positive psychologists conduct large-scale surveys of what people strive for most in life, parenting is rarely at the top of the list.

Unfortunately, this sort of thinking is prominent. For instance, Boorse (2011) argued that "for physiology, the highest level goals, of the organism as a whole, are individual survival and reproduction" (p. 27). However, as DeYoung and Krueger (2018) pointed out, while it's true that these two outcomes—survival and reproduction—are relevant to understanding evolutionary function, they are not necessarily the organism's highest-level goals. Goals involve representations of the desired future state, and must actually be represented within the system in some way. This is critical for guiding behavior and interpreting ongoing experience. There are many people who have the goal of having sexual intercourse with someone they are attracted to, but who nonetheless have no interest in reproduction. Therefore, relying strictly on an evolutionary definition of function doesn't necessarily tell us the full story about the full gamut of human capacities and strivings.

Consider Wakefield's (2007) definition of healthy mental function as those processes that were selected by evolution and that facilitated propagation of genes over generations. It's an alluring next step to then equate all trait manifestations that are *not* in accord with evolutionary function as necessarily dysfunctional and therefore detrimental to a rich life. However, I believe we must resist that urge. Equally as important as evolutionary mismatches are individual mismatches. Many instances of languishing are the result of a society that is a poor fit to one's unique strengths (Fromm, 1960). What may appear to be an evolutionary mismatch could, on closer reflection, be an indication that the society is thwarting someone's personal goal (Kaufman, 2013).

Human beings, like every other living organism on this planet, are *cybernetic systems* (e.g., Carver & Scheier, 1998; DeYoung, 2015; Wiener, 1961). As cybernetic systems, we have multiple, often conflicting, goals. For each goal, we have representations of where we currently are in relation to the goal, and we have a set of operators that (it is hoped) allow us to move closer to the goal. Evolutionary and cybernetic functions often overlap; indeed, many of the goals of the cybernetic systems exist precisely because they increased reproductive fitness over the course of evolution. However, cybernetic and evolutionary functions can *diverge*, and we see this most clearly and most often in human beings, who "have evolved an unprecedented degree of flexibility in the goals we can adopt" (DeYoung & Krueger, 2018).

There are many examples of humans behaving in ways seemingly contrary to reproduction and even survival. DeYoung and Krueger (2018) gave the examples of hunger strikes and celibacy, for instance. Now, it may turn out that these behaviors do ultimately serve an evolutionary-selected function, so I certainly don't want to imply that every behavior has to be conducive to fitness in the present to be evolutionarily functional. My point is merely that understanding the cybernetic function of an individual, in addition to understanding its potential evolutionary function, helps us come to a deeper understanding of human flourishing and well-being (also see DeYoung, 2015). Yes, the sorts of problems that present themselves on the clinician's couch often overlap with evolutionary dysfunction—problems with survival, mating, and parenting, for instance—but some personal problems do not have an obvious evolutionary function, and the best choice for living the good life is not always the answer that was selected for by evolution.

For instance, the humanistic psychologists, which provided the foundational philosophy behind positive psychology, emphasized the inherent conflicts that come with existing as a human being (see Maslow, 1962; Rogers, 1961). Among many things that stand out among humans, we are a *self-aware* ape. We are an ape concerned with personal identity, creative expression, and purpose. Let's not underestimate the unique manifestations of the exploration drive that exists among humans, or downplay the extraordinary implications of the many ways this drive can express itself among humans, for living a good life. Other apes don't have existential crises. There's a reason why we do, and a reason why this matters.

In *The Sane Society*, the social psychologist and humanistic philosopher Erich Fromm acknowledged our biological drives but argued that the "human condition" involves the fundamental tension between our common nature with other animals and our uniquely developed capacities for self-awareness, reason, and imagination. As Fromm noted,

> The problem of [human] existence, then, is unique in the whole of nature: [we have] fallen out of nature, as it were, and [are] still in it; [we are] partly divine, partly animal; partly infinite, partly finite. The necessity to find ever-new solutions for the contradictions in [our] existence, to find ever-higher forms of unity with nature, [our fellow human] and [ourself] is the source of all psychic forces which motivate [us], of all [our] passions, affects and anxieties.

Similarly, the existential psychotherapist Irvin Yalom (1980) argued there are four "givens of existence" that humans must reconcile: (a) *death*, the inherent tension between wanting to continue to exist and the inevitability of perishing; (b) *freedom*, the inherent conflict between the seeming randomness of the universe and the heavy burden of responsibility that comes with the freedom to choose one's own destiny; (c) *isolation*, the inherent tension between, on the one hand, wanting to connect deeply and profoundly with other human beings and be part of a larger whole and, on the other hand, never fully being able to do so, always remaining existentially alone; and (d) *meaninglessness*, the tension between being thrown into an indifferent universe that has no inherent meaning and yet wanting to find some sort of purpose for our own individual existence in the incomprehensibly short time in which we are living on this planet.

In summary, I believe there are three things that evolutionary psychologists can do to help elucidate the good life. The first is to identify evolutionary mismatches that we can rectify, as Geher and Wedberg so accurately point out in this book. Second, I believe it's important to consider meaning and purpose, and the higher-level goals that are unique to each individual, and give people a sense of truly existing on this planet. Third, I recommend identifying evolved tendencies (e.g., the desire to cheat on a spouse) that can get in the way of our well-being. Coming to a more complete understanding of why these evolved self-undermining tendencies are so strong could be a useful insight for those striving for something more enduring.

I say enduring because humans are truly unique in the long timescale of our goals and in our flexibility to choose which goals we most wish to prioritize. One promising method for reaching clarity in our goals and becoming less immediately reactive to our evolved instincts is the practice of meditation, which has received increasing research attention in positive psychology (e.g., Wright, 2017). As Robert Wright has written about so convincingly in *Why Buddhism Is True*, the disciplined practice of meditation, over time, can help one to become less slavish to our evolved instincts and be kinder, gentler, and happier as a result. It can also help one witness more beauty in the world.

In my view, those working within the field of positive evolutionary psychology should look not only at the individual parts that may have increased reproductive fitness in our distant past, but also at the *whole person*, right here, right now, listening to their dreams, desires, priorities, and conflicts and helping them become something *greater* than the sum of their parts. In other words, this exciting new field can help people become more fully human.

Scott Barry Kaufman
Philadelphia, PA
November, 2017

References

Baumeister, R. F., Vohs, K. D., Aaker, J. L., & Garbinsky, E. N. (2013). Some key differences between a happy life and a meaningful life. *The Journal of Positive Psychology, 8,* 505–516.

Boorse, C. (2011). Concepts of health and disease. In F. Gifford (Ed.), *Handbook of the philosophy of science, vol. 16: Philosophy of Medicine* (pp. 13–64). New York: Elsevier.

Buss, D. (2014). *Evolutionary psychology: The new science of the mind.* London: Psychology Press.

Carver, C., & Scheier, M. (1998). *On the self-regulation of behavior.* New York: Cambridge University Press.

DeYoung, C. G. (2015). Cybernetic big five theory. *Journal of Research in Personality, 56,* 33–58.

DeYoung, C. G., & Krueger, R. F. (2018). A cybernetic theory of psychopathology. *Psychological Inquiry, 29,* 117–138.

Dunn, D. S. (Eds.). (2017). *Positive psychology: Established and emerging issues.* Abingdon, England: Routledge.

Fromm, E. (1960). *The sane society.* New York: Holt, Rinehart, & Winston.

Himmelfarb, G. (1996). *Darwin and the Darwinian revolution.* Chicago: Dee. (Original work published 1959)

Ilardi, S. S. (2010). *The depression cure: The 6-step program to beat depression without drugs.* Boston: Da Capo Lifelong Books.

Kaufman, S. B. (2013). *Ungifted: Intelligence redefined.* New York: Basic Books.

Keltner, D. (2009). *Born to be good: The science of a meaningful life.* New York: Norton.

Kenrick, D. T., & Griskevicius, V. (2013). *The rational animal: How evolution made us smarter than we think.* New York: Basic Books.

Kenrick, D. T., Griskevicius, V., Neuberg, S., & Schaller, M. (2010). Renovating the pyramid of needs: Contemporary extensions built upon ancient foundations. *Perspectives on Psychological Science, 5,* 292–314.

Kurzban, R. (2012). *Why everyone (else) is a hypocrite: Evolution and the modular mind.* Princeton, NJ: Princeton University Press.

Lopez, S. J., Pedrotti, J. T., & Snyder, C. R. (2014). *Positive psychology: The scientific and practical explorations of human strengths.* Thousand Oaks, CA: Sage.

Maslow, A. (1962). *Toward a psychology of being.* New York, NY: Wiley.

Peterson, C. (2006). *A primer in positive psychology.* New York: Oxford University Press.

Rogers, C. (1961). *On becoming a person: A therapist's view of psychotherapy.* New York, NY: Mariner Books.

Seligman, M. E. P. (2012). *Flourish: A visionary new understanding of happiness and well-being.* New York: Atria Books.

Vaillant, G. (2009). *Spiritual evolution: How we are wired for faith, hope, and love.* New York: Harmony.

Wakefield, J. C. (2007). The concept of mental disorder: Diagnostic implications of the harmful dysfunction analysis. *World Psychiatry, 6,* 149–156.

Wiener, N. (1961). *Cybernetics—Or control and communication in the animal and the machine* (2nd ed.). New York: MIT Press/Wiley.

Wright, R. (2017). *Why Buddhism is true.* New York: Simon & Schuster.

Yalom, I. (1980). *Existential psychotherapy.* New York: Basic Books.

Scott Barry Kaufman is the former science director of The Imagination Institute, an instructor of a course in positive psychology at the University of Pennsylvania, and author of *Ungifted* and *Wired to Create.*

Preface

Homicide. Jealousy. Conflict. Deception. These are but some of the topics that have been famously studied by evolutionary psychologists over the past few years. The field of evolutionary psychology has developed something of a reputation for studying the dark side of the human experience. And some even see the field as somehow endorsing dark, nasty, selfish behavior.

After teaching evolutionary psychology for years, I honestly had enough! Sure, evolutionary psychologists do study all those dark topics. But you know, better understanding things like homicide and conflict actually can help us reduce these phenomena. Further, some folks might be surprised to learn that evolutionary psychologists also study a broad suite of phenomena that represent the bright side of life.

Evolutionary psychologists are actually known to study such topics as love, gratitude, happiness, and cooperation. These are all phenomena that are usually placed under the umbrella of *positive psychology*. But guess what? Not only do evolutionary psychologists also study these bright features of our lives. Armed with Darwin's famously powerful intellectual toolkit, evolutionary psychologists regularly shed important new light on our understanding of all these phenomena.

Take gratitude as an example. Gratitude has been extensively studied by positive psychologists, with much work showing that experiencing and expressing gratitude is essentially good for the soul and leads to all kinds of positive emotional and social outcomes. That's great. And I think that gratitude is fully deserving of being amplified in all of our lives.

However, what many folks do not know is that gratitude is also studied, extensively, by evolutionary psychologists. And this work (including much of the research conducted by Michael McCullough, for instance) uses Darwinian principles as a framework to help us understand the ultimate factors that underlie gratitude.

Evolutionary psychologists inherently study all kinds of phenomena from a stepped-back approach. With gratitude, as the working example, this means asking questions such as *why gratitude exists in the first place, what might be evolutionarily adaptive about experiencing and expressing gratitude,* as well as *what proximate factors might exist to facilitate or inhibit the experience of gratitude.*

This intellectual approach has proven to be extremely powerful in understanding all kind of behavioral phenomena. Evolutionary psychologists who study the positive features of the human experience are, whether they think about it or not, advancing the goals of positive psychology, which, generally speaking, seeks to better understand factors that help people experience the good life.

It's time for the world to see the positive light that radiates from work in the evolutionary behavioral sciences. And it's time for positive psychology to connect with evolutionary psychology and develop new ways of studying the positives of the human experience that fully and importantly take Darwinian ideas into account.

The world is ready for the emergence of *positive evolutionary psychology.* And here, among the pages of this book, is where you will find it.

I hope you enjoy this book and I hope it helps you to better understand the world as well as our place in it. And I hope that these ideas help to shed all kinds of new light for you in terms of what it means to live the good life.

We all have a ticket on the same ride—here's to getting the most out of it along the way. Together.

Genuinely, Glenn Geher

Acknowledgments

GG: This book, like any great human product, is the result of a community of individuals providing support, wittingly or unwittingly, in various kinds of ways. Our original editor, Abby Gross at Oxford, is a truly exceptional person to work with–showing vision, compassion, and diligence in her working with us on this project. To have the reins turned over to Nadina Persaud has been such a gift, as Nadina follows Abby's footsteps when it comes to being supportive and visionary. And Katharine Pratt's boots-on-the-ground assistance with this project has been nothing short of stellar over the years. Oxford University Press is a class act and we are thankful for this fact. In my academic world, several of my scholarly friends have contributed to the ideas in this book via conversations and dialog over the years. Scott Barry Kaufman is clearly among those who has helped shaped the vision of this book–Scott is a five-star person in all regards and his support and friendship has helped shape this book in many positive ways. At SUNY New Paltz, I am lucky to be surrounded by a swarm of amazing students–many of whom have contributed positively to this book in many ways. We like to call ourselves the New Paltz Evolutionary Psychology Lab–and working with these students is at the center of my work–and they inspire me every day. Students who have played particularly crucial roles in helping to shape the ideas in this book include: Amanda Baroni, Kian Betancourt, Jacqueline Di Santo, Morgan Gleason, Rich Holler, Olivia Jewell, Brianna McQuade, Eden Nitza, Vania Rolon, Miriana Ruel, Stephanie Stewart-Hill, and Gratia Sullivan. Keep up the great work, students of the NP EP Lab–you will do great things moving forward. I have no doubt. Beyond these students I must give special thanks to lab alumni sine qua non, and co-author, Nicole Wedberg–who proved to me and to the world when she was a student that hard work, resilience, resourcefulness, and a smile will always eventually lead to great successes of all kinds. Thanks, Nicole, for being the inspiration that you are. Keep up the great work. And I'd be remiss if I did not thank my amazing life partner, Shannon. Shannon's boundless support and warmth have helped my thinking about the good things in life in all kinds of beautiful ways–many of which have directly affected the ideas in this book. Shannon is, simply, the sunshine of my life–and for her support, I am deeply grateful. Finally, I thank those special kids in my world–Megan and Andrew–who both make me smile every day and who, in their own unique ways, are grabbing this world by the horns already. Megan and Andrew: Stay awesome and keep on being you! The world is already a better place because of your being in it. Stay the course. And stay positive.

SECTION I
POSITIVE EVOLUTIONARY PSYCHOLOGY DEFINED

Simply put, positive evolutionary psychology is the marriage of the fields of *evolutionary psychology* and *positive psychology*. Each of these areas of psychology has been of great interest to students and scholars over the past several decades. Each has turned up in textbooks, university-level courses, graduate programs, and extensive research that have helped advance our understanding of the human condition.

But there is something of a problem. In academia, it is often the case that separate areas of inquiry, even if closely related, progress independently from one another. This has been the case with the fields of evolutionary and positive psychology. Each of these fields, both with important insights into who we are, has heretofore progressed as if the other does not exist.

As evolutionary psychologists, we see this problem as deep, largely because we believe that evolutionary psychology, with its focus on how Darwinian principles can shed light onto all aspects of human behavior, has a great deal of untapped potential when it comes to illuminating the positive aspects of psychological life. And illuminating the positive aspects of psychological life is precisely the core goal of positive psychology.

Thus, positive evolutionary psychology is an approach to scholarship that utilizes the mountain of work in the evolutionary behavioral sciences to shed light on the positive features of psychological, social, and community life in humans. As you'll see throughout this book, we believe that this approach has the capacity to advance the field of positive psychology by leaps and bounds. Further, you will see that this approach has a broad array of implications regarding how you personally can live a richer life.

1

What Is Positive Evolutionary Psychology?

Forgiveness. Love. Family. Giving. Helping. Childrearing. Spirituality. Community. Each of these concepts pertains to positive aspects of life. And each has been illuminated by the evolutionary approach to understanding human behavior.

This book seeks to document the marriage of two of the most substantial trends in modern psychology: *evolutionary psychology*, which focuses on the use of evolutionary principles to help us better understand human behavior (see Geher, 2014), and *positive psychology* (see Gable & Haidt, 2005), which focuses on the capacity for human growth and other positive aspects of the human experience.

From where we stand, the discipline of psychology is fully ready for the emergence of *positive evolutionary psychology*. These fields have both led to a great deal of excitement among researchers and students of the behavioral sciences in recent years. And they both have a demonstrated record of providing new insights into what it means to be human. Further, as we see it, these two approaches to psychology are fully compatible with one another. Laying the groundwork for the basic assumptions and parameters of positive evolutionary psychology, then, should have the capacity to lead to new research questions and new insights into the positive aspects of the human experience.

Goals of This Work

This book is ambitious in scope. It seeks to develop a new area within the field of psychology by merging the fields of evolutionary and positive psychology. The rationale for such a goal is straightforward. Evolutionary psychology, informed fully by Darwin's great insights into the nature of life, has proven to be a powerful approach to understanding the full suite of human behaviors. Evolutionary psychology has shed light onto such topics as the nature of emotions (see Ekman & Friesen, 1986); attraction between individuals (see Gallup & Frederick, 2010); warfare (see Smith, 2009); altruism (see Burnstein, Crandall, & Kitayama, 1994);

and more. Given its focus on evolutionarily fundamental questions regarding *why* human behavioral patterns exist in the first place, evolutionary psychology has demonstrated an unparalleled capacity to generate new research questions and, as a result, new research findings. And this translates into new things that we now know about what it means to be human.

Positive psychology is also leaving a significant mark in terms of our understanding of human psychology. An approach that inherently seeks to advance our understanding of positive aspects of the human experience, positive psychology has excited the minds of many researchers and students, leading to new insights into such topics as happiness (see Haidt, 2006; Jayawickreme, Forgeard, & Seligman, 2012; Seligman, 2011); creativity (see Kaufman & Gregoire, 2015); and growth, as opposed to stress, after traumatic experiences (see Calhoun & Tedeschi, 1999). With a straightforward charge of focusing our science on the positive aspects of the human experience, positive psychology is making great advances in helping us uncover the secrets of human greatness.

Scholars of human behavior are, of course, also passengers in the ride of life. Behavioral scientists find themselves in the position of studying topics that all humans, including themselves, have a strong interest in on a personal level. Evolutionary psychology provides us a powerful set of tools to understand any aspect of human behavior. Positive psychology provides us with a framework for understanding and helping to cultivate positive aspects of the human experience. As behavioral scientists with both (a) a strong background in using evolutionary principles to inform research and (b) an interest in helping effect positive outcomes for people in general, we believe that carving out the details of positive evolutionary psychology is an important step in the development of psychology as a discipline.

To begin to understand how the fields of evolutionary psychology and positive evolutionary psychology can be integrated into a single, powerful framework, we need to first understand the basic elements of each of these approaches to psychology. We turn to descriptions of the basics of these areas next.

What Is Evolutionary Psychology?

In its simplest form, evolutionary psychology is the idea that human behavior is part of the natural world. That is to say that the same processes that govern all phenomena in the natural world, shaped by evolutionary forces, also ultimately govern human behavior.

In short, evolutionary psychologists work to explain and understand behavioral patterns and psychological processes in terms of Darwinian principles, such as natural selection. For instance, consider the fear of snakes, which is a very common fear found in people across cultures (Öhman, & Mineka, 2001). An evolutionary approach to understanding this phenomenon focuses on how such a fear may be the result of evolutionary forces. Can we understand fear of snakes as a product of natural selection? And what exactly would such an explanation look like?

Natural selection is pretty much exactly what it sounds like: It's the idea that some attributes of organisms are selected to exist into the future by nature. If some quality is helpful in getting an organism to survive or reproduce, then it *naturally* is more likely to exist into the future compared with alternative qualities.

So let's go back to our example regarding fear of snakes. Imagine one of our ancestors living in sub-Saharan Africa, where snakebites account for a significant proportion of human deaths. Now picture him walking around the jungle without a care in the world, and when he hears a slithering sound on the trail behind him, he doesn't even flinch. This carefree attitude may help make him a likable fellow in his tribe, but guess what? He's also likely to die an untimely death at the fangs of a venomous snake.

Now picture a vigilant member of the tribe who's always on the lookout for danger when roaming the jungle. His strongest concern is a potential run-in with a venomous snake. He can't quite describe why, but he is deathly afraid of any and all snakes—generalizing to serpentine stimuli of any kind. Well this guy might be perceived as sort of uptight by his fellow tribe members, but you know what? At least he's alive!

So we can think of fear of snakes as a psychological feature that was *selected naturally* or *selected by nature.* Those ancestors of ours who had this feature were naturally more likely to survive and reproduce compared with their non–snake-fearing counterparts, who were (naturally) more likely to experience premature mortality.

From an evolutionary perspective, fear of snakes can be understood as a classic behavioral adaptation—a psychological feature that likely allowed our ancestors who possessed this feature to have an increased likelihood of survival or reproduction. And to the extent that this psychological feature may have even a small heritable (genetic) component, those ancestors who possessed this feature were likely, after surviving and mating, to pass on this feature (fear of snakes) to offspring.

The Power of Evolutionary Psychology

In addition to benefiting from the clear logic that underlies a Darwinian approach to understanding the nature of life, evolutionary psychology has demonstrated itself to be a powerful framework for understanding human behavior, leading to novel and significant insights into many facets of what it means to be human.

In a now-classic paper on the powerful nature of the evolutionary perspective in psychology, Ketelaar and Ellis (2000) argued that a good scientific framework is one that (a) leads to new questions and, as a result, (b) leads to new understanding and insight. Evolutionary psychologists (see Geher, 2014) study behavior, seeking to apply evolutionary principles to help us best understand why different behavioral phenomena exist. The field of evolutionary psychology is full of new findings about human behavior that we simply would not have without the mountain of research that this field has cultivated. Here are three of the biggies—things we now know about human behavior thanks to evolutionary psychology:

1. **Men are more than twice as likely to experience early mortality (death) during young adulthood compared with women (Kruger & Nesse, 2006).**
   Men are more likely to die than are women at any and all phases of the life cycle. Applying an evolutionary lens, Kruger and Nesse (2006) hypothesized that this phenomenon should be exacerbated during young adulthood when males are more likely to be courting mates and, as a result, engaging in male/male (intrasexual) competition. And that's exactly what they found.

2. **Stepparents, compared with genetic/biological parents, are (by a large order of magnitude) more likely to engage in filicide (killing of offspring) (see Daly & Wilson, 2005).**
   Filicide is universally seen as horrific. So it would benefit humanity writ large to understand its antecedents. Applying evolutionary-based reasoning, Daly and Wilson (2005) reasoned that as stepparents do not share

the same genetic investment with offspring as biological parents do, then stepparents might be more likely to engage in filicide. And this is, by a large order of magnitude, exactly what they found.

3. **Across all reaches of the world, men, compared with women, show a stronger preference for variety in mates (see Schmitt et al., 2003).**
    Schmitt and his collaborators hypothesized that across multiple human groups, males would show a stronger preference for variety of mating partners compared with females (as there are fewer evolutionary costs for males in mating with multiple partners compared with the costs for females). Based on one of the world's largest and most diverse human research samples ever studied, these researchers provided compelling evidence to support their evolution-based hypothesis: Across the world (and across methods of measurement), males demonstrated a stronger preference for variety in sexual partners compared with females.

In fact, these three examples represent only the tip of the iceberg. The field of evolutionary psychology has proven to be remarkably powerful in the past few decades, shedding light on all facets of the human condition.

Basic Principles of Evolutionary Psychology

Given the centrality of the evolutionary psychological perspective in making the case for a new field of positive evolutionary psychology, it's important for us to demarcate several of the basic concepts of the field here. This section includes a brief summary of some of the core ideas that underlie this basic field of inquiry.

Natural Selection Applies to Behavior

Observable qualities of organisms with some heritable component are *naturally selected* if they lead to probabilistic increases in survival or reproductive success of the organism. Behavioral patterns in humans, which are often partly heritable, may be naturally selected. Evolutionary psychologists thus often conceptualize behavioral patterns as qualities of humans that have the capacity to increase the likelihood of survival or reproduction.

The Importance of Human Universals

Human universals play an important role in evolutionary psychology, as universals (e.g., the *fear of snakes* example described previously; see Öhman & Mineka, 2001) likely evolved as adaptations to help our ancestors survive or reproduce. Otherwise, these things wouldn't be universal. So if some behavioral trait is universal across human populations, that's a clue that it's a product of basic evolutionary forces such as natural selection.

Evolutionary Mismatch

The lion's share of human psychology evolved over eons in the African Savanna. And only recently (in evolutionary time) have humans lived in large-scale postagricultural and postindustrial societies. Since organic evolutionary processes take a long time to effect change, our minds are actually better suited to ancestral, preagricultural contexts than they are to modern contexts. Many features of modern living (e.g., the large-scale availability of processed foods or the fact that many people do not live near extended family members) do not match our evolved psychology—and problems (such as physical or mental health issues) are often the result.

Strategic Pluralism

Human behavior is highly flexible, and many different behavioral strategies have evolved in our species across various psychological domains (see Gangestad & Simpson, 2000). For instance, under highly stable conditions, humans are more likely to pursue long-term mating strategies (e.g., monogamy) than they are under relatively unstable conditions (see Figueredo et al., 2005). The evolutionary psychological approach seeks to understand the nature of behavioral flexibility in terms of evolutionary principles.

Multiple Evolutionary Forces Underlie All Human Behavior

Natural selection is but one evolutionary force that underlies human behavior. Darwin documented several other evolutionary forces, such as *sexual selection*, which exists when some trait increases the probability of reproductive success, often at a cost to likelihood of survival. Further, cultural forces have been shown to result from evolutionary processes, and such cultural evolution plays a significant role in shaping human behavior. Evolutionary psychologists consider the panoply of evolutionary forces that work together to produce behavioral outcomes.

Both Ultimate and Proximate Factors Underlie All Behavior

Any evolved psychological feature has both *ultimate* causes, which speak to the adaptive function of the feature and address how this feature allowed our ancestors to survive or reproduce, and *proximate* causes, which correspond to the immediate factors that underlie some behavior (see Tinbergen, 1963). These proximate causes are like the nuts and bolts that allow some psychological or behavioral process to play out. For instance, from an ultimate perspective, aggressive behavior in males has its ultimate roots in the fact that male ancestors of ours who were successful in physical competitions with other males were more likely to survive and reproduce. From a proximate perspective, on the other hand, males tend to have relatively high free-floating levels of testosterone, and this hormone works as a catalyst that leads to aggressive acts.

Evolutionary Psychology in a Nutshell

In a nutshell, evolutionary psychology is the application of evolutionary principles to the study of psychological phenomena. The most straightforward approach to evolutionary psychology pertains to understanding psychological features as *adaptations* shaped by natural selection—features of our psychology that have some heritable component and that helped confer survival or reproductive advantages to our ancestors.

The evolutionary approach to psychology has been wildly successful over the past several decades, leading to new insights into such aspects of humanity as religion, education, politics, health—and more. As you'll see, the purpose of this book, in carving out the field of positive evolutionary psychology, is to join the field of evolutionary psychology with the emerging field of positive psychology, with an eye toward capitalizing on the power of evolutionary psychology to help illuminate such positive areas of human functioning as altruism, love, creativity, and the cultivation of well-functioning human communities.

What Is Positive Psychology?

When Martin Seligman took over as the president of the American Psychological Association several years ago, he put out a famous call to action for a new field of study (see Seligman, 2011). The field of psychology had gone full guns into a particular direction, and he didn't like it. In short, Seligman criticized the field of psychology for focusing on the negatives: *psychopathology, abnormality, anxiety*, etc. There are wonderful things about humans that often get eclipsed by this large focus on the negatives connected with our psychological lives. *Enough already!* In a highly effective manner, Seligman put out a call for work in the field of positive psychology. He suggested that we as scholars in the behavioral sciences should focus our energies not only on what goes wrong with human functioning but also on what we are doing right. What are the factors that are associated with flourishing? How can we increase human productivity and happiness? How can we facilitate a sense of meaning among people? What factors lead to positively functioning human communities?

In the words of Christopher Peterson (2008), positive psychology "the scientific study of what makes life most worth living." A common metaphor to help put a face to this idea focuses on how to get individuals from a 0 to +5 instead of focusing on how to get people from -5 to 0. Gable and Haidt (2005, p. 1) have defined positive psychology as "the study of the conditions and processes that contribute to the flourishing or optimal functioning of people, groups, and institutions." Given the goals of positive psychology, this field is quite applied in nature, often including research into various cognitive and behavioral strategies designed to lead to positive outcomes across the human experience.

The Power of Positive Psychology

Positive psychology is becoming widely recognized as an important and powerful field. There is now a considerable amount of evidence-based publications demonstrating the power of positive psychology in helping to uncover new information about the positive aspects of the human experience. Three research examples from the field of positive psychology that exemplify the core principles of this field are presented next.

Positive Psychology Interventions Can Lead to Overall Increases in Well-Being

A *positive psychological intervention* (PPI) can be simply described as a type of intervention that focuses on increasing positive feelings and behaviors in long-term manner (Schueller & Parks, 2014). In a sense, improving such positive affective states sits at the core of much of the work in this field. Multiple meta-analyses (or studies that examine and summarize data from multiple other studies) that have examined the efficacy of PPIs designed to increase positive affect in a long-term manner have shown that PPIs can lead to overall moderate increases in well-being along with decreases in depressive symptoms (Boiler et al., 2013; Sin & Lyubomirsky, 2009). On this point, Seligman, Steen, Park, and Peterson (2005) have demonstrated, in fact, that certain PPIs have the potential to increase happiness while decreasing depressive symptoms over extended periods of time. Positive psychological approaches to behavioral change work, and they can work over the long haul.

Positive Psychology-Based Applications Can Lead to Improved Functioning Across Various Kinds of Environments

Given its charge of trying to improve the human condition across a broad array of areas, work in the field of positive psychology has been conducted across several kinds of human contexts. For instance, in a large-scale study of the potential for PPIs to work in home environments, Sergeant and Mongrain (2015) implemented a long-term online PPI program related to the home lives of several thousand adults. Among distressed individuals, decreases in depressive symptoms and increases in life satisfaction were documented, and these effects were found to persist even 6 months after completing the program (see Sergeant & Mongrain, 2015). Importantly, this research utilized an Internet-based approach which allowed people to ex-perience PPI from their own homes, implicating that online PPIs may be a feasible option for many, and they may be effective across a broad suite of human contexts.

A Positive Psychology-Based Approach Can Provide Useful Guidance to People of All Ages

The positive psychology approach is not limited to adults. PPIs have led to increases in hope, the ability to handle social stress, courage, humanity, and love among emerging adults (Leontopoulou, 2015). Adolescents in school have also benefited from PPIs. School-based PPIs, designed to en-hance the mental health and well-being of students, have been effective in strengthening self-efficacy and optimism while reducing general distress, anxiety, and depression symptoms (Shoshani & Steinmetz, 2013). A positive psychology approach has a place to be useful throughout the life span.

Basic Principles of Positive Psychology

There are myriad topics that are studied within the field of positive psychology, and one could argue that we have only scratched the surface of true understanding when it comes to human flourishing. Areas of study include optimism, love, forgiveness, awe, hope, curiosity, laughter, the psychobiological mechanisms of happiness and morality, mindfulness meditation, well-being therapy, and many others that are aimed at improving well-being. Scott Barry Kaufman, who worked for years as the director of research for the Center for Positive Psychology at the University of Pennsylvania (known as the epicenter of the field of positive psychology across the globe), writes on positive psychology wherein he describes the field in terms of both *experiences* and *character traits*. Based partly on his conception of the field, here are three foundational principles of positive psychology:

A Focus on Positive Affective States (Past, Present, and Future) Can Lead to Improved Functioning

Positive experiences may exist in the past tense, including such feelings as nostalgia, pride, and satisfaction. They may also correspond to *current* feelings such as pleasure, contentment, and laughter. Even future-based feelings, such as optimism and hope, are included in the full suite of positive experiences. Desired psychological outcomes such as decreased depressive symptoms and increased well-being apply directly to experience across the past, present, and future (Boiler et al., 2013; Sin & Lyubomirsky, 2009).

Cultivating Character Traits With a Positive Orientation Can Help Advance Well-Being

There are many positive character traits that are examined in the field of positive psychology. Such traits include wisdom, humility, temperance, resilience, courage, compassion, love, justice, self-efficacy, and creativity among others. Positive character traits are important as they help to set the parameters for how individuals will approach growth opportunities. For instance, several research findings have found that hope and zest are character traits that strongly predict life satisfaction and perseverance in patients with traumatic brain injury. Courage and self-regulation are other positive traits that are associated with overall well-being (Hanks, Rapport, Waldron-Perrine, & Millis, 2014).

On this point, Jonathan Haidt (2006) talked about the importance of teaching students to focus on harnessing their existing strengths or character traits as they strive to develop positively, rather than trying to "fix" their weaknesses. Often, our own strengths can help us get out of a jam or solve a problem. The positive psychology approach seeks to explore these (and other) approaches to character traits in efforts to help understand the antecedents of human flourishing.

An Understanding of Positive Aspects of Mind Can Be Applied Across All Life Domains

One of the unique and accessible aspects of positive psychology is its broad applicability. Like evolutionary psychology, positive psychology is highly interdisciplinary and has a place for all parts of life, arguably. The implications of research and practice in positive psychology are widely advantageous. To explicate this point, consider the fact that positive psychology has been useful for addressing such a broad array of psychological phenomena as alleviating overall distressed individuals of depressive symptoms (Sergeant & Mongrain, 2015) and improving financial planning (Asebedo & Seay, 2015). It is used in the clinical setting (Duckworth, Steen, & Seligman, 2005); to enhance relationships within households (Schueller & Parks, 2014); to help members of minority groups to fight social inequality (Domínguez, Bobele, Coppock, & Peña, 2015); to aid the long-term unemployed (Dambrun & Dubuy, 2014); and to help teens navigate the pitfalls associated with development (Leontopoulou, 2015; Shoshani & Steinmetz, 2013). In fact, Kaczmarek et al. (2017) have shown that increased public smiling, facilitated by a positive psychological approach, leads to great success among scientists! The field of positive psychology is increasingly demonstrating itself as a highly useful approach to help address the many facets of our lives.

Positive Psychology in a Nutshell

Essentially, positive psychology is an applied discipline dedicated to focusing on the more positive aspects of human behavior. It is a field of study just as much as it is an approach to methods designed to enhance the lived human experience. Positive psychology embodies any focus or methodology for increasing overall happiness, well-being, and all of the feelings, traits, and experiences therein.

While the targeted scientific study of ultimate fulfillment and human flourishing may be relatively new, the experience of and striving toward these important human goals is actually ancient. Across human evolutionary history, positive emotional states such as happiness and love have consistently provided motivation for thousands of generations of humans. This book focuses on merging the subareas of positive psychology with the broad-based field of evolutionary psychology, with an eye toward addressing how the selection for so many behaviors may be useful in guiding us to a richer life.

The Marriage of Evolutionary and Positive Psychology

Every now and again, two distinct academic fields join forces, leading to a powerful new field that takes advantage of the benefits of each of the fields that feed into it. As an example, consider the field of cognitive neuroscience. This field is now a major area of inquiry, but it wasn't always. Decades ago, there were two pretty independent areas of psychological inquiry: *cognitive psychology* and *physiological psychology*. Cognitive psychology, which largely emerged in the 1960s, focuses on such cognitive processes as mental imagery, text comprehension, memory, and logic processing. Physiological psychology, also with roots in the middle of the prior century, focuses on the physical qualities that correspond to psychological and behavioral outcomes. Largely, physiological psychologists study the nervous system and how different aspects of the nervous system, such as sections of the brain, particular neurotransmitters, or hormones, come to effect psychological or behavioral outcomes.

The field of cognitive neuroscience, now represented by college courses bearing this title at colleges and universities all around the world, is essentially a marriage of cognitive psychology and physiological psychology. At some point, researchers realized that the methods used by physiological psychologists (such as the use of EEG [electroencephalographic] technology to study electrical activity of the brain) could be applied to research questions asked by cognitive psychologists (e.g., addressing how an adult learns words in a new language) to lead to insights into how cognitive processes are connected with brain processes. This exciting area of inquiry is regularly leading to new information regarding how neural processes and cognitive processes connect with one another.

The primary point of this book is to provide a road map for a field created in a similar fashion—by the marriage of two existing fields of academic inquiry.

For the purposes of thinking about the marriage of evolutionary and positive psychology, then, consider these brief conceptions of each of these fields independently:

- **Evolutionary psychology** is essentially an approach to understanding any aspects of human behavior via the large-scale application of the most powerful set of ideas that exists in the life sciences: evolution.
- **Positive psychology** is essentially an area of psychological inquiry that focuses on the positive aspects of the human experience, such as human flourishing, creativity, and the capacity for love.

Given these conceptions of these two areas, then, it's clear that these fields are not at all incongruous with one another. Evolutionary psychology is the application of a set of tools to understanding any area of psychology. Positive psychology is the use of scientific psychology to shed light particularly on the positive aspects of the human experience.

So why not use the powerful set of tools that comprise evolutionary psychology, then, to help us better understand the positive aspects of the human experience?

The Scope of This Book

The 11 chapters of this book are divided into four broad sections. This first section introduces the basic ideas of positive evolutionary psychology in a broad form and includes a full chapter on the topic of *evolutionary mismatch*, or situations in which ancestral conditions do not match modern environments, as this concept underlies many of the basic questions asked by evolutionary psychologists.

A second section essentially is the meat and potatoes of the book, demarcating the basic content areas to be elucidated by positive evolutionary psychology. Such areas include politics, religion, love, social relationships, resilience, and an evolutionary perspective on taking a positive approach to life.

The third section focuses on what we might call *applied positive evolutionary psychology*. This section focuses on how work in this field can help shed light on such important issues as human health and on the building of well-functioning communities.

The fourth and final section has a single chapter designed to demarcate the implications from the rest of this book for both future research in the behavioral sciences and the implications for how to live the good life starting today.

If our work in writing this book is successful, then readers should walk away with a deep understanding of how the evolutionary perspective in psychology can illuminate the positive aspects of the human experience.

Acknowledgments

Some content from this chapter was adapted from Glenn Geher's (2015) *Psychology Today* blog post "The Power of Evolutionary Psychology." Other material was adapted from his (2015) post "Bigger Than Ourselves." Glenn owns the copyright to the material.

References

Asebedo, S. D., & Seay, M. C. (2015). From functioning to flourishing: Applying positive psychology to financial planning. *Journal of Financial Planning, 28*(11), 50–58.

Boiler, L., Haverman, M., Westerhof, G. J., Riper, H., Smit, F., & Bohlmeijer, E. (2013). Positive psychology interventions: A meta-analysis of randomized controlled studies. *BMC Public Health, 13*, 119.

Burnstein, E., Crandall, C., & Kitayama, S. (1994). Some neo-Darwinian decision rules for altruism weighing cues for inclusive fitness as a function of the biological importance of the decision. *Journal of Personality and Social Psychology, 67*(5), 733–789.

Calhoun, L. G., & Tedeschi, R. G. (1999). *Facilitating posttraumatic growth: A clinician's guide.* New York: Routledge.

Daly, M., & Wilson, M. (2005). The "Cinderella effect" is no fairy tale. *Trends in Cognitive Sciences, 9*(11), 507–508.

Dambrun, M., & Dubuy, A. (2014). A positive psychology intervention among long-term unemployed people and its effects on psychological distress and well-being. *Journal of Employment Counseling, 51*(2), 75–88.

Domínguez, D. G., Bobele, M., Coppock, J., & Peña, E. (2015). LGBTQ relationally based positive psychology: An inclusive and systemic framework. *Psychological Services, 12*(2), 177–185.

Duckworth, A. L., Steen, T. S., & Seligman, M. E. (2005). Positive psychology in clinical practice. *Annual Review of Clinical Psychology, 1*, 629–651.

Ekman, P., & Friesen, W. V. (1986). A new pan cultural facial expression of emotion. *Motivation and Emotion, 10*, 159–168.

Figueredo, A. J., Vásquez, G., Brumbach, B. H., Sefcek, J. A., Kirsner, B. R., & Jacobs, W. J. (2005). The K-factor: Individual differences in life history strategy. *Personality and Individual Differences, 39*(8), 1349–1360.

Gable, S. L., & Haidt, J. (2005). What (and why) is positive psychology? *Review of General Psychology, 9*(2), 103–110.

Gallup, G., & Frederick, D. A. (2010). The science of sex appeal: An evolutionary perspective. *Review of General Psychology, 14*, 240–250.

Gangestad, S. W., & Simpson, J. A. (2000). The evolution of human mating: Trade-offs and strategic pluralism. *Behavioral and Brain Sciences, 23*, 573–644.

Geher, G. (2014). *Evolutionary psychology 101*. New York: Springer.

Geher, G. (2015). Bigger than ourselves. *Psychology Today* blog post.

Geher, G. (2015). The power of evolutionary psychology. *Psychology Today* blog post.

Haidt, J. (2006). *The happiness hypothesis*. New York: Basic Books.

Hanks, R. A., Rapport, L. J., Waldron-Perrine, B., & Millis, S. R. (2014). Role of character strengths in outcome after mild complicated to severe traumatic brain injury: A positive psychology study. *Archives of Physical Medicine and Rehabilitation, 95*(11), 2096–2102.

Jayawickreme, E., Forgeard, M. J., & Seligman, M. E. (2012). The engine of well-being. *Review of General Psychology, 16*(4), 327–342.

Kaczmarek, D. L., Behnke, M., Kashdan, T. B., Kusiak, A., Marzec, K., Mistrzak, M., & Włodarczyk, M. (2017). Smile intensity in social networking profile photographs is related to greater scientific achievements. *The Journal of Positive Psychology, 13*(5), 435–439. doi:10.1080/17439760.2017.1326519

Kaufman, S. B., & Gregoire, C. (2015). *Wired to create: Unraveling the mysteries of the creative mind*. New York: Penguin Books.

Ketelaar, T., & Ellis, B. J. (2000). Are evolutionary explanations unfalsifiable? Evolutionary psychology and the Lakatosian philosophy of science. *Psychological Inquiry, 11*, 1–21.

Kruger, D. J., & Nesse, R. M. (2006). An evolutionary life history understanding of sex differences in human mortality rates. *Human Nature, 74*(1), 74–97.

Leontopoulou, S. (2015). A positive psychology intervention with emerging adults. *The European Journal of Counselling Psychology, 3*(2), 113–136.

Öhman, A., & Mineka, S. (2001). Fears, phobias, and preparedness: Toward an evolved module of fear and fear learning. *Psychological Review, 108*(3), 483–522.

Peterson, C. (2008, May 16). What is positive psychology, and what is it not [Web log post]? *Psychology Today* blog. Retrieved from https://www.psychologytoday.com/us/blog/the-good-life/200805/what-is-positive-psychology-and-what-is-it-not

Schmitt, D. P., Alcalay, L., Allik, J., Ault, L., Austers, I., Bennett, K. L., . . . Zupanèiè, A. (2003). Universal sex differences in the desire for sexual variety: Tests from 52 nations, 6 continents, and 13 islands. *Journal of Personality and Social Psychology, 85*, 85–104.

Schueller, S. M., & Parks, A. C. (2014). The science of self-help: Translating positive psychology research into increased individual happiness. *European Psychologist, 19*(2), 145–155.

Seligman, M. E. P. (2011). *Flourish: A visionary new understanding of happiness and well-being*. New York: Atria Books.

Seligman, M. E. P., Steen, T. A., Park, N., & Peterson, C. (2005). Positive psychology in progress. Empirical validation of interventions. *American Psychologist, 60*, 410–421.

Sergeant, S., & Mongrain, M. (2015). Distressed users report a better response to online positive psychology interventions than nondistressed users. *Canadian Psychology*, *56*(3), 322–331.

Shoshani, A., & Steinmetz, S. (2013). Positive psychology at school: A school-based intervention to promote adolescents' mental health and well-being. *Journal of Happiness Studies*, *15*(6), 1289–1311.

Sin, N. L., & Lyubomirsky, S. (2009). Enhancing well-being and alleviating depressive symptoms with positive psychological interventions: A practice-friendly meta-analysis. *Journal of Clinical Psychology*, *65*, 467–487.

Smith, D. L. (2009). *The most dangerous animal*. New York: St. Marten's Griffin.

Tinbergen, N. (1963). On aims and methods of ethology. *Zeitschrift für Tierpsychologie*, *20*(4), 410–433.

2

Examples of Positive
Evolutionary Psychology

In Chapter 1, we demarcated what positive evolutionary psychology is. To help carve the way for the remainder of this book, here we provide three examples of what positive evolutionary psychology looks like—to help put a face to this area of inquiry. To demonstrate how multifaceted positive evolutionary psychology is, we include examples that vary quite a bit from one another in terms of content. Specifically, our examples address (a) the experience of love and (b) the importance of reciprocity on social behaviors.

The Evolutionary Psychology of Love

Positive psychology focuses largely on understanding and fostering positive emotional experiences. And when it comes to such experiences, it's hard to top the emotional experience of love. Love, a cross-cultural universal (see Fisher, 1993), has been studied extensively by evolutionary behavioral scientists, leading to important insights into its origins and functions.

An enormous body of work on the evolutionary psychology of love from the past several decades (e.g., Fisher, 1993) has demonstrated how strong our love for another can be. Love, an inherently selfless act, then, is a foundational part of the human evolutionary story. While strict monogamy is rarely the norm in societies found around the world, mating systems with a monogamous component are quite common. Long-term mating, which corresponds to mating systems that are at least partially monogamous, emerge in species in which the young are helpless and essentially need multiple adults to help them survive and reproduce. Many species of songbirds provide great examples.

Newly hatched robins, for instance, grow to full fledglings that fly out of the nest within a few weeks. They metabolize extraordinarily during this period of development, and they need very much in the way of worms and other food sources to be able to grow in an optimal manner. One parent is simply not going to be able to get enough worms for the growing birds. As such, multiple adults are needed, and this is exactly the kind of situation that tends to often facilitate the evolution of a long-term mating system. And wouldn't you know it: A pair of robins tends to be monogamous throughout an entire season, and when they come back north the next year, there is even a good chance that they will find one another again and rekindle things.

Humans are similar to robins in this regard. Our young are also very altricial—they need tons of help in early development—and a single adult who has multiple responsibilities is not likely to be as effective at providing all the needed resources compared with a pair or group of adults. As with robins, this fact regarding the high levels of parental investment needed in our species has led to variants of monogamous mating. A pair of adults is much better positioned, all things equal, to be able to provide the resources needed for a young growing child compared with what could be provided by a single parent.

With this backdrop in mind, Helen Fisher (1993) carved out an evolutionarily informed model of love, focusing on the adaptive nature of romantic love in creating a pair bond that is best situated to provide needed resources for a developing altricial child. Love, which includes the proximate glue of increased oxytocin levels in the brain, helps members of a couple stay strongly connected and aligned with one another, caring about the interests of the other in a deep and genuine way. A relationship built on a foundation of love provides a great context for that all-important domain of life: childrearing.

There are few goals in life that are as positive as achieving truly loving relationships. If you want to understand what love is and why it exists in the first place, you'd be best off starting with an evolutionary perspective.

The Evolutionary Psychology of Reciprocity in Relationships

Humans are a communal ape, and much of our evolved psychology bears on this point. Sometimes people seem to think that because the evolutionary perspective has a focus on *The Selfish Gene* (see Dawkins, 1976/1989), or qualities that evolved to benefit individuals themselves at a cost to others, they mistakenly think that evolutionary psychology is essentially the psychology of selfish and self-centered behavior. In fact, once you scratch the surface a bit, you find that nothing could be further from the truth.

A core part of the evolutionary psychology of emotions and social behaviors pertains to the *moral emotions* as they relate to a deep history of *reciprocal altruism* (R. L. Trivers, 1971) in our species. In certain species, individuals help conspecifics (members of their same species) that are not related to themselves along kin lines. In some species, reciprocal altruism has evolved. This is essentially a kind of helping in which you help someone with some kind of expectation of help in return. Across time, individuals develop long-standing mutually beneficial relationships based partly on a foundation of reciprocal altruism. For instance, you may have a neighbor (let's call him Ed) who has a tall ladder that he always lets you borrow whenever the need strikes. And you might have a great leaf blower that you leave under your deck—and that you make available to Ed to use at any time.

As master evolutionary biologist Robert Trivers taught us with his classic work on this topic from the 1970s, a key to understanding relationships in humans pertains to the moral emotions (see R. Trivers, 1985), a set of emotions that evolved specifically to help optimize reciprocal altruism in one's world. For instance, consider guilt. Suppose that you borrowed Ed's ladder a few weeks ago, but your leaf blower is out of service and you never bothered to tell Ed about this. Then he comes over to borrow it, only to get shut out. How does that make you feel? In a situation like this, you'd probably feel a splash of guilt.

According to Trivers, guilt evolved to help motivate us to engage in *relationship-restoring behaviors* (e.g., apologetic behaviors) to ultimately maintain the highly beneficial social relationships that we may have (unwittingly, perhaps) put into jeopardy. So you feel guilt regarding your failed leaf blower—you apologize to good old Ed—and then you fix your leaf blower the next day and send Ed a text that is kind of like this: "Ed—leaf blower is all fixed—plz use whenever."

This little text will likely lead to a little text from Ed, such as, "Thanks so much. All good."

Helping others and forming long-standing relationships based on reciprocal altruism is a core feature of what it means to be human. To understand the nature of positive interpersonal relationships with a bird's-eye view, we need an evolutionary perspective.

The Evolutionary Psychology of Gratitude

Another piece of the human social–emotional puzzle that corresponds to reciprocal altruism has to do with the expression of gratitude. In fostering relationships with others, an unconscious goal seems to be to get others to see you as someone who is likely to reciprocate altruism—someone who should be looked to as a *friend* and *ally* in a sea of individuals. In short, it's good to radiate to potential friends the following message: *You can trust me—I will be there for you. If you need something, you just let me know.*

One social–emotional process that helps convey this information pertains to the expression of gratitude. When someone does something for us, he or she has taken time and resources away from him- or herself to do something for us. From an evolutionary perspective, this is huge and needs to always be acknowledged. Time that I spent baking you cookies is time that I could have been watching Netflix and eating Cheetos.

Expressions of gratitude ("Thank you so much!" "I really appreciate it!" "You have no idea how much this means to me!" etc.) play an important role in helping keep people connected to one another. They help make the person who is expressing gratitude a likely candidate for reciprocating altruistic acts in the future.

Imagine this scenario: It's December and you're feeling kind of jolly. Before you head to work on Monday, you decide to wake up early and bake some chocolate chippers for some of your colleagues. You drop off a box with your coworker Sally, who stands right up at her desk, looks at these cookies, and emphatically says, "Oh you shouldn't have! These look lovely! This must have taken you so much time! Thank you so much for this wonderful holiday treat—I can't wait to share them with my family!"

All right then! So now you head down the hall and you find Theresa, the next on your list for dropping the cookies off. Theresa's response is a bit less gracious (i.e., *less infused with gratitude*). She doesn't stand up, and she barely looks at you. Her response is, "What are you doing—trying to give me a heart attack with this cr*p?" And that's that.

Well now, let's think about Sally and Theresa in terms of longer term social relationships. First, based simply on the information presented here, who do you think is more likely to do something good for you at some future point? I'm thinking Sally; her over-the-top expression of gratitude is a marker or signal that she can be counted on in your social circle and can be counted on to provide you with some helpful benefits in your future. We take expressions of gratitude that way.

Second, think about how each of their responses made you feel. Sally probably made you feel great; her response made the time that went into those cookies well worth it. Theresa, on the other hand, made you feel kind of icky—nothing positive there. She kind of made you feel like a fool. The feelings that others in our world lead us to experience play a major role in our choices of social partners and in group membership.

As a result of our long-standing evolutionary history of reciprocal altruism, moral emotional responses, such as expressions of gratitude, play an important role in shaping positive psychological experiences. Positive psychologists who focus on strategies to increase positive emotional experiences can benefit extraordinarily from an understanding of the evolutionary psychology that underlies such emotional states.

A Kinder, Gentler Evolutionary Psychology

From where we stand, the large-scale incorporation of evolutionary principles into the behavioral sciences has improved all areas of our field, and there is an enormous amount of untapped potential. This said, evolutionary psychology has not always been well received among all academic circles. One of the goals of the development of positive evolutionary psychology as a field is to create a context for scholars and students to benefit from the powerful tools of the evolutionary perspective while concurrently focusing on research questions that inherently bear on the positive aspects of being human.

With the publication of David Buss's (1999) textbook, *Evolutionary Psychology*, a field was born. Granted, the ideas of this field have stewed in the minds of academics since at least the mid-1800s, but Buss's seminal work gave the field the push it needed at the time to enter *full-blown* status.

From the outset, Buss described evolutionary psychology as a broad field of inquiry, focusing on all aspects of behavior from the perspective that behavioral patterns are the result of evolutionary forces. This said, as a prolific researcher, Buss's own work in the field focused largely on issues of human mating, and much of that particular work has focused largely on issues of evolved behavioral sex differences. Buss should never be faulted for being too prolific of a researcher, by the way. But as we see it, his mountain of research shaped perceptions of the field—unwittingly. You see, many people outside evolutionary psychology started to conflate evolutionary psychology with the study of evolved behavioral sex differences. Of course, sex differences research is, conceptually, only a slice of evolutionary psychology, but, again, owing largely to the massive research program of Buss and his colleagues, many other academics saw evolutionary psychology and the study of evolved behavioral sex differences as one and the same.

Of course, none of this should be a problem, but there's a catch. It turns out that many people have a strong resistance to accepting the idea of evolved behavioral sex differences (see Geher & Gambacorta, 2010). The kinds of people who are so resistant to this idea tended to be people who

- Are very politically liberal
- Are academics (especially in the social sciences)
- Have no children

These features are often characteristic of academics at many universities and colleges across the nation. So you can see that if evolutionary psychology is being conflated with the idea of evolved behavioral sex differences, and academics generally don't like that idea, then these same academics might not like evolutionary psychology.

Interestingly, this same research (Geher & Gambacorta, 2010) found that, generally, academics had no problem with other facets of evolutionary psychology. For instance, the participants in the study were fine acknowledging that evolutionary explanations helped us understand the behavior of non-human animals such as dogs and cats. Further, they acknowledged the utility of evolutionary psychology in helping us understand non–mating-related phenomena such as

- Fear of heights
- Disgust at rotten food
- The universal nature of human emotional expression

The implications of this research are consistent with the goals of this book. Research on the politics of evolutionary psychology has shown that resistance to this field is often largely political—and comes often from academics who don't accept biological explanations of male/female differences.

The movement toward positive evolutionary psychology, demarcated in this book, moves beyond the political resistance to evolutionary psychology within academic circles. Instead of suggesting that evolutionary psychologists conduct more carefully designed research to try to convince critics of the utility of evolutionary psychology, the approach presented here suggests simply moving forward. Darwin's ideas on the evolution of life, which fully include behavior, are simply as powerful a set of ideas as can be seen in any area of academia. To reject these ideas as applied to issues of human behavior is, simply, a major mistake. Rejecting the inclusion of evolutionary principles in the behavioral sciences is just as misguided as the idea of rejecting evolution as an explanation for life itself.

The approach taken in this book suggests that a subset of future research in the field of evolutionary psychology explicitly address issues that are central to the positive aspects of living. The advancement of this subset of evolutionary psychology proper, which we call positive evolutionary psychology, will inherently sidestep prior political struggles.

As demonstrated in the beginning of this chapter section regarding examples of positive evolutionary psychology, focusing on love, social reciprocity, and gratitude, we can see that work in this area is already well under way. We believe that a field explicitly focused on understanding positive human experiences from an evolutionary perspective will have the potential to continue such great work in a way that circumvents political problems that have followed past efforts to integrate Darwin's ideas into the behavioral sciences.

References

Buss, D. M. (1999). *Evolutionary psychology: The new science of the mind.* New York: Allyn & Bacon.

Dawkins, R. (1989). *The selfish gene.* Oxford: Oxford University Press. (Original work published 1976)

Fisher, H. (1993). *Anatomy of love—A natural history of mating, marriage, and why we stray.* New York: Ballantine Books.

Geher, G., & Gambacorta, D. (2010). Evolution is not relevant to sex differences in humans because I want it that way! Evidence for the politicization of human evolutionary psychology. *EvoS Journal: The Journal of the Evolutionary Studies Consortium, 2*(1), 32–47.

Trivers, R. (1985). *Social evolution.* Menlo Park, CA: Benjamin/Cummings.

Trivers, R. L. (1971). The evolution of reciprocal altruism. *Quarterly Review of Biology, 46,* 35–57.

3

Ape Out of Water

Evolutionary Mismatch and the Nature of Who We Are

In 2012, after being home to various classes of primates for more than 100 years, the renowned Monkey House at the Bronx Zoo closed. The zoo's current primate inhabitants reside in spaces designed to better match their natural habitats. The shift is part of a broader movement in zoos around the world, based on the highly reasonable idea that all species have evolved to fit particular environmental conditions. Such features that typify an organism's ancestral conditions characterize its *environment of evolutionary adaptedness* or EEA (Bowlby, 1969). Housing a monkey, whose ancestors go millions of years deep into specific African jungle environments, in a small cage in the zoo of a large city was simply *evolutionarily misguided*—and arguably cruel.

Animals essentially *need* to have many key features of their ancestral environments in their current conditions because their bodies and evolved psychological processes are the result of evolutionary forces that took place under these specific conditions. In an evolutionarily novel and unnatural environment, solid research showed that various primates will demonstrate signs of physiological and psychological stress (see Harlow & Suomi, 1971).

In short, a monkey in a cage goes bananas! And this metaphor is captured perfectly in the visionary work of Kurt Vonnegut (1968), per his book of short stories simply titled *Welcome to the Monkey House*. This metaphor resonates so strongly with an evolution-based approach to understanding modern humans because, as we see it, the joke is this: Humans in modern environments are, essentially, like monkeys in cages at the zoo. As you will see in a broad array of examples in this book, our modern environments are mismatched from the conditions that our bodies and minds were shaped to experience in many ways. And just as Harlow's monkeys had a hard time living outside their natural ecological contexts, the same goes for modern humans.

A key principle of evolutionary psychology (see Geher, 2014) is the notion that modern humans in Westernized societies experience important instances of evolutionary mismatch. Understanding this point is essential in understanding much of what it means to be human today (see Giphart & Van Vugt, 2016). Further, understanding evolutionary mismatch is critical to an understanding of positive evolutionary psychology.

Positive psychology seeks to determine the conditions under which people thrive—the factors that lead to such outcomes as greatness and achievement along with positive social and emotional outcomes. Understanding evolutionary mismatch is, then, absolutely critical for understanding all the basic goals of positive psychology. If you want to understand what factors lead to happiness, then you have to understand what kinds of outcomes under ancestral conditions led to feelings of positive affect—outcomes such as positive relationship outcomes, success in obtaining resources, success in the lives of kin (such as the birth of a niece or a nephew), and so forth. Further, if you want to understand factors that lead to problems in the modern day, you can easily look to instances of mismatch between modern conditions and ancestral conditions. People are obese because modern food offerings (often comprising highly processed ingredients and filled with sugars and starches that did not exist in these same proportions under ancestral conditions) take advantage of our evolved food preferences without considering how damaging the large-scale availability of cookies and cheeseburgers can be. People in big cities are often anxious, lonely, and depressed because our ancestors always lived only in small communities, communities in which they had many long-standing social and familial connections. In so many ways, our modern environments are discordant from ancestral human conditions.

If we as scholars into the human condition expect to advance the goals of positive psychology, we had darn well better understand the implications of evolutionary mismatch as it relates.

Ten Ways That Your Life Is Mismatched to Ancestral Conditions

To put a face to the many ways that evolutionary mismatch affects our everyday lives, consider the following 10 instances of mismatch that, often unbeknown to us, affect our mental health on a daily basis.

1. You are surrounded in your day-to-day life by a higher proportion of strangers than would ever have been true of our preagrarian hominid ancestors.

Before the advent of agriculture, our ancestors always lived in small, no-madic clans, clans that rarely exceeded 150 individuals (see Dunbar, 1992). You interacted with these people during your entire lifetime, and you were related to about half of them. Did you ever notice that you get nervous giving a talk in front of a group of strangers? Well, guess what? This very common occurrence partly relates to the fact that doing something like that would have never happened under ancestral conditions.

2. You run into a higher total number of people each day than our preagrarian hominid ancestors ever would have.

Dunbar's number of 150, is, put simply, itty-bitty by any modern standards. Imagine you are driving across the United States and come on a sign for a town; the sign simply says this: East Bumblestead, KS, Population 150. You might laugh. You might take a picture. You might say to your driving buddy something like, "Wow! People actually live in such a tiny place!? They must be so backward!"

Well, in fact, based on extensive research in the field of evolutionary studies, the folks in East Bumblestead, Kansas, may well have it going on much more than you realize. In small communities, crime is relatively rare, as are mental health problems (see Srivastava, 2009). And when people all know one another well, they have all kinds of motivation to be nice to one another; after all, they can, with good reason, expect to see one another again and again.

My town of New Paltz is considered a "small town"—and our popu-lation is over 14,000—a size that exceeds Dunbar's 150 exponentially. Now think about someone living in New York City or in Chongqing, China, with a population of more than 30,000,000. There is a reason that there are benefits to small-scale living—and the key is found in our evolutionary past.

3. You have the option of spending 90% of your waking
hours sitting at a desk—and you often *exercise* this option.

Modern biological anthropologists estimate that people in a modern nomadic group might migrate, with a combination of walking and running, about 20 miles in a typical day. Think about that! If you are a typical
American, then you probably walk about 1–2 miles a day. You probably have
a nice, comfortable bed. You probably get up and drive to work (in a nice and
comfortable car). You probably have a comfortable chair in your office area,
and you may well sit in that chair for about 30 hours a week. When you come
home, you likely spend a good bit of quality time on your couch, watching
others experience life on a large screen that is across from you.

Perhaps this is a bit too negative. After all, if you are reading this book, you
probably are interested in taking care of yourself. You may well belong to a
gym and run, swim, or cycle for a hobby. This said, think about how incredibly mismatched our lives are from ancestral conditions in terms of exercise.
Our preagrarian ancestors had no choice but to exercise a lot each and every
day. We now have to pay pricey gym memberships to get ourselves to exercise,
with attending the gym often being something of a chore for many people.

Interestingly, positive psychologists who have explored the psychological
benefits of exercise (e.g., Wright & Zhao, 2018) have consistently found mental
health benefits such as increases in feelings of self-worth and well-being along
with reductions in signs of depression and anxiety. These proximate outcomes
associated with modern exercise regimens speak to the fact that our minds
and bodies seem best adapted to conditions in which we exercise a lot.

Want to understand evolutionary mismatch as it relates to modern human
living? You don't have to think much beyond exercise.

4. Your extended family includes people dispersed
across hundreds or thousands of miles (think
between New York and Florida).

Under ancestral social conditions, family was close by. When someone had a
baby, the entire community was there to help. In fact, research across a broad

array of primates, including humans (see Hrdy, 2009), showed that child-rearing in pre-Westernized societies almost always takes the form of a village of women. Across varied pre-Westernized societies, childrearing largely falls on the mother, but the duties are shared with her mother—and her sisters—and with many of the other women in her community. Sarah Blaffer Hrdy called this the *mothers and others* approach to childrearing, and, with no offense meant toward men, this approach is considered the basic and natural approach to childrearing in our species.

Imagine growing up like that. A basic feature of such an ecological context is the fact that there are always lots of genetically related kin around to help. From an evolutionary perspective, this is actually quite significant. Kin, after all, are people in the world with whom you share a genetic interest (see Hamilton, 1964). As has been documented in various species, humans demonstrate multiple forms of nepotism: We are biased toward helping our genetic kin. This makes good evolutionary sense, as helping kin is, in effect, helping your genes as they exist in the bodies of others.

In short, in natural ecological conditions, humans tend to be surrounded by kin, and this reality starts before conception. Being isolated from kin in modern-day contexts has the capacity to wreak havoc on our emotional states.

5. You were raised in some variant of a nuclear family.

A corollary to the prior point regarding the diaspora of extended family pertains to the fact that a high proportion of people now are raised in the confines of a nuclear family. Raising young offspring is, without question, one of the most time- and energy-intensive features of the human life span. Human babies are deeply altricial, needing a great deal of parental effort to survive and ultimately to thrive. As Sarah Hrdy (2009) pointed out, in pre-Westernized groups around the globe, women typically secure the help of other women in raising young offspring.

Fast-forward to North America in the twenty-first century. Here, the nuclear family reigns. In this modern context, people are often isolated, living in urbanized settings in dual-income households. In such a scenario, Mom and Dad often have few opportunities to secure nonpaid help with childrearing. From an evolutionary perspective, this can be a problem.

6. You have been exposed to more images of violence than ever would have been possible for preagrarian hominids.

Research on violent images is staggering. By the time a modern human is an adult, he or she has seen thousands of instances of violence on TV, on the Internet, etc. Worse, there is consistent research in the social sciences demonstrating that viewing such violence has the capacity to predispose people to violent acts themselves.

Violent acts, and markers of such acts, such as bloodshed, are of intrigue to the human mind (see Bildhauer, 2013). To put a face to this fact, think about the thousands of horror movies that have been produced. Under ancestral conditions, seeing violence and bloodshed never took place remotely. If you experienced violence and bloodshed, that meant that you were in the thick of it—and it would have been terrifying because of the obvious fitness-/survival-related implications.

As with many classes of stimuli, humans like to make *supernormal stimuli* (Tinbergen, 1953). That is, we artificially create extreme versions of stimuli that we would have evolved to respond to in some way. Horror movies and action movies that are riddled with violence are essentially this: human-made supernormal stimuli that exploit the fact that violence was emotionally powerful (if not positive) under ancestral conditions.

7. You were likely educated in an age-stratified system— spending each of several years in a group comprising about 25 others who matched you in age—being taught in a classroom environment by a few specially designated "teachers." You likely spent a lot of time sitting behind desks in the process.

Evolutionary psychologist Peter Gray of Boston College has famously documented how highly mismatched our modern educational systems are from what would have surely been typical of ancestral conditions on this front (see Gray, 2011; Gruskin & Geher, 2018). In a large-scale and cross-cultural anthropological analysis of educational systems in pre-Westernized societies, Gray found that nothing resembling modern education in places like the United States was found anywhere. In pre-Westernized contexts, there is often not a distinct concept of *education*. Children go out during the day, and they play with one another. They are loosely supervised by older children. They learn the trades of the community from one another in that context.

In no cultures that Gray and his colleagues studied did children spend 8 hours a day in groups of kids who were all the same age as one another—learning almost exclusively from one adult woman. They weren't sitting at desks and working largely on projects to develop skills that were only tangentially related to basic survival-based aspects of life.

Gray observed that boys in pre-Westernized cultures, in particular, were regularly outside, running around. This is a far cry from sitting at a desk all day, and it does not take too much to think about how this relates to the fact that boys are diagnosed with attentional issues by about 10 times the rates compared with girls in modern educational contexts. Want to know why so many kids are diagnosed with attention deficit hyperactivity disorder these days? Part of it, to be sure, is the result of evolutionary mismatch.

In fact, a study conducted by Gruskin and Geher (2018) found that college students who reported having had a high proportion of evolutionarily relevant educational experiences early on (such as having multiage learning experiences in elementary school, having elementary experiences that focused on free play, etc.) scored as enjoying school more and as performing better than their peers academically. The nod toward an evolutionarily natural approach to learning is not just lip service.

8. You are exposed regularly to politics at a global scale, often discussing and being involved in issues that potentially pertain to thousands, millions, or even billions of other humans.

No matter where you land on the political spectrum, you probably think that the world is a mess right now. The 2016 election in the United States was one of the most divisive political events in history, and it has had global ramifications. People have been marching in the streets by the hundreds of thousands—across the world—in the name of such issues as women's rights, immigrants' rights, healthcare, education, science, and more. Concomitant with this trend toward activism is the fact that nations around the world are in conflict with one another at uneasy levels.

From an evolutionary perspective, it's actually not that difficult to understand why large-scale politics are a mess. Under ancestral conditions, our nomadic ancestors never dealt with large-scale politics. Politics during ancestral conditions would have all been small-scale, local politics. Clans comprised about 150 individuals, and people encountered the same individuals across their life spans. There was no such thing as a municipality, a county, a state, or a nation. There certainly was nothing comparable to nations competing for resources with other nations, and this mismatch may well help us to understand why large-scale politics are such a mess right now.

9. You spend a great deal of time interacting with *screens* and *devices*—having the evolutionarily unprecedented possibility of almost never having to be bored at all.

All human social interactions under ancestral conditions took place in direct, face-to-face contexts. For the lion's share of human evolution, there was no such thing as a phone, mail, email, Facebook, etc. If you wanted to have a social interaction, you needed to go out and talk with someone. If you wanted to date, there was no Tinder. You had to do it the old-fashioned way with a face-to-face interaction, and probably a whole lot of nerves.

These days, a large proportion of our interactions takes place in virtual, electronic ways (see Holler, 2017). Dating is done via websites and apps, and people can create dating profiles, where they can edit and re-edit their photos and descriptions for hours before posting them. Think about how different this process is from when you have to meet a potential mate in a face-to-face context! Further, most online interactions are strictly visual, while face-to-face interactions include audio, tactile, and olfactory stimuli as well.

The next time you are out and about, check out how many people are looking at their phones and other devices. These days, anywhere you go— from the subway to the airport, from restaurants to movie theaters, from baseball games to bridal showers—people are constantly interacting with their phones. Sure, there are benefits; there is something to be said for having a small electronic gadget in your pocket that can answer almost any question in the world instantaneously—while playing your favorite song—but there are costs as well. Cell phones are famously addictive, and they play a high-profile role when it comes to wreaking havoc on family time in households across the world.

From an evolutionary perspective, the appeal of cell phones becomes apparent. We evolved to desire instant gratification. We evolved to be psychologically rewarded for connecting socially with others. We evolved to desire all kinds of things that we can quickly pull up on our cell phones anywhere at any time. Like so many modern technologies, cell phones exploit our evolved psychology. And as is usually the case when a technology exploits our evolved psychology, there are costs that tag along with the benefits.

To address this emergent problem, some positive psychologists have developed positive psychological interventions to reduce screen use (Khazaei, Khazaei, & Ghanbari-H, 2017). Consistent with the evolutionary perspective presented here, this intervention was found to increase positive emotional experiences and the quality of social relationships.

10. You can eat an entire diet of processed foods, and you live in a world in which processed foods are cheaper and more accessible than natural foods.

Perhaps the most basic and well-documented mismatch in modern humans pertains to diet (see Giphart & Van Vugt, 2016). Our ancestors in the African savanna not only often ran into drought and famine, but also rarely had access to foods that were high in sugar and fat. Across evolutionary time, our ancestors evolved taste preferences such that they would prefer, and thus seek, foods high in sugar and fat content—precisely because such foods were rare and adaptive under those ancestral conditions. Well, look at us now. We now have highly processed foods that are full of sugar and fat, and these foods are, ironically, incredibly cheap and easy to access. What a mismatch! And if you are looking for the cause of modern obesity problems (and resultant health issues such as Type 2 diabetes) that characterize so many Western nations, look no further. This evolutionary mismatch may well be the ultimate cause of all of these problems (see Wolf, 2010).

After owning a deep fryer for a few months a few years ago, I (G. G.) put on a bunch of weight (yes, I am blaming the deep fryer!). Three events got me to think very hard about adopting a natural approach to diet. First, I ran an intensive Spartan race with my brother and a friend. They were both in peak shape as they took on this challenge. I, on the other hand, almost died. A few weeks later, I hiked up Mt. Washington in New Hampshire with some old friends. I was the guy at the end of the pack who barely made it to the top. About a week later, I had my annual checkup with my doctor. He politely called me fat—and that was it! As an evolutionist, I knew exactly what do. From that point, I changed my diet immediately and fully to an all-natural set of foods: fruits, vegetables, and meat. In 3 months, I got back to my high school weight, and I have remained there ever since. Want to lose weight? Consider the evolutionary history of the human body.

Bottom Line

This list is certainly incomplete and, at best, preliminary; there are undoubtedly many other worthy contenders. That said, we are hopeful that this list can help open the eyes of those interested in human psychology to the importance of evolutionary mismatch in understanding all aspects of who we are. Further, we hope that this list makes it quite clear just how important an evolutionary approach is to shedding light on how we can redirect humans to be on a thriving trajectory during life. Positive psychology, the psychology that focuses on cultivating humans in a way that gets them to thrive, can only realize its potential with an evolutionary framework in hand.

As demonstrated across this chapter, modern life is mismatched in countless ways from the conditions that characterized the environments of our ancestors during the bulk of human evolution. Such evolutionary mismatches often lead to problems of body, society, and mind. Fortunately, when it comes to ways to address problems of evolutionary mismatch, evolutionary psychology provides clear answers (see Geher, 2014). Want guidance for how to live a richer and healthier life? Want to bring the field of positive psychology to the next level? Take a look at what Darwin had to say.

And welcome to the monkey house.

Acknowledgments

Some content from this chapter was adapted from Glenn Geher's (2015) *Psychology Today* blog post, "Evolutionary Mismatch and What You Can Do About It" as well as his (2013) post, "10 Ways That Our Lives Are Out of Whack." Glenn owns the copyright to the material.

References

Bildhauer, B. (2013). Medieval European conceptions of blood: Truth and human integrity. *Journal of the Royal Anthropological Institute, 19*, 57–76.

Bowlby, J. (1969). *Attachment and loss. Vol. 1. Attachment.* New York: Basic Books.

Dunbar, R. I. M. (1992). Neocortex size as a constraint on group size in primates. *Journal of Human Evolution, 22*(6), 469–493.

Geher, G. (2013). Ten ways that our lives are out of whack. *Psychology Today* blog.

Geher, G. (2014). *Evolutionary psychology 101.* New York: Springer.

Geher, G. (2015). Evolutionary mismatch and what you can do about it. *Psychology Today* blog.

Geher, G., Carmen, R., Guitar, A., Gangemi, B., Sancak Aydin, G., & Shimkus, A. (2016). The evolutionary psychology of small-scale versus large-scale politics: Ancestral conditions did not include large-scale politics. *European Journal of Social Psychology, 46*(3), 369–376. doi:10.1002/ejsp.2158

Giphart, R., & Van Vugt, M. (2016). *Mismatch.* London: Robinson.

Gray, P. (2011). The special value of age-mixed play. *American Journal of Play, 3*, 500–522.

Gruskin, K., & Geher, G. (2018). The evolved classroom: Using evolutionary theory to inform elementary pedagogy. *Evolutionary Behavioral Sciences, 12*, 336–347.

Hamilton, W. D. (1964). The genetical evolution of social behaviour. I. *Journal of Theoretical Biology, 7*, 1–16.

Harlow, H. F., & Suomi, S. J. (1971). Social recovery by isolation-reared monkeys. *Proceedings of the National Academy of Sciences of the United States of America, 68*, 1534–1538.

Holler, R. R. (2017). *Friends, love, and tinder.* (Unpublished master's thesis, State University of New York at New Paltz).

Hrdy, S. B. (2009). *Mothers and others: The evolutionary origins of mutual understanding.* Cambridge, MA: Harvard University.

Khazaei, F., Khazaei, O., & Ghanbari-H, B. (2017). Positive psychology interventions for Internet addiction treatment. *Computers in Human Behavior, 72*, 304–311.

Srivastava, K. (2009). Urbanization and mental health. *Industrial Psychiatry Journal, 18*, 75–76.

Tinbergen, N. (1953). *The herring gull's world.* London: Collins.

Vonnegut, K. (1968). *Welcome to the monkey house.* New York: Delacorte Press.

Wolf, R. (2010). *The paleo solution.* Las Vegas, NV: Victory Belt.

Wright, V. J., & Zhao, E. (2018). *Masterful care of the aging athlete* (pp. 25–29). New York: Cham Springer.

SECTION II
DOMAINS OF POSITIVE EVOLUTIONARY PSYCHOLOGY

Areas within an academic discipline focus on various content domains. Social psychology has the areas of social cognition, social influence, and attitude formation, for example. Developmental psychology includes cognitive development, social development, and moral development. Personality psychology includes work on trait taxonomies, personality development, and unconscious personality processes.

Positive evolutionary psychology includes all areas of psychology where the evolutionary perspective has the capacity to inform positive aspects of human psychological life. A subset of such areas is included in this second section of the book. This section thus essentially comprises the meat and potatoes of the book.

Content areas included here focus on political psychology, religion and spirituality, altruism and the moral emotions, happiness, social interactions, love, relationships, parenting, dealing with hardships, and resilience—addressed across five chapters. Generally, this section focuses on specific findings from the field of evolutionary psychology that have direct implications for living the good life.

4

The Political Ape

The Psychology of the Greater Good

So if you're a student of human nature, here's a puzzle for you: Where did democracy come from? It certainly doesn't exist across all corners of the animal kingdom. Lions don't have democracy. The strongest and most dominant male gets access to all of the females and dibs on any and all other resources. The same is true for silverback gorillas and a variety of nonhuman primates.

Yet we see egalitarianism and democratic kinds of systems in various human groups. On the other hand, we see many nondemocratic systems, such as totalitarianism, in human groups across the globe as well. How can an evolutionary approach to understanding humans inform the nature of human politics and governance systems?

The Expansion of Human Groups or The Great Evolutionary Leap

Let's begin by thinking about the origins of systems of governance in humans. Early hominids lived in relatively small groups compared with the groups that we live in now. Before the advent of agriculture 10,000 years ago, no humans lived in cities. No humans lived in but a single location, surrounded by thousands of others—all with their own specialized jobs. Ancestral humans were generalists, and they lived in small groups.

This said, at some point in evolutionary history, *Homo sapiens* made what David Sloan Wilson (2007) called a *great evolutionary leap*. We expanded our in-groups to include non-kin. In other words, at some point, a clan of *Homo sapiens* emerged and included individuals from multiple family units—all working in a coordinated fashion toward common goals, all working together as a single unit.

As we discussed throughout this book, the impact of this great evolutionary leap cannot be overemphasized when it comes to understanding the human condition. Consider the following:

- Neanderthals had brains that were actually larger than the brains of *Homo sapiens*, yet fossil evidence suggests that their social worlds never expanded much beyond their immediate kin. They (for most functional purposes) bit the evolutionary dust (see Geher et al., 2017).
- Human religions across the globe underscore connections among people who share the same belief systems—across kin lines. Religious groups that aggressively recruit members from all walks of life tend to be the religious groups that succeed. And religious groups largely focus on helping others from the same group (see Wilson, 2007)—across kin lines.
- With the expansion of human groups to include non-kin, human group size expanded. We now live in groups that number into the millions (think the United States) or even billions (think China).
- A large and coordinated group of individuals who see themselves as *on the same team* has the capacity to do things that no individual human, no matter how smart or how strong, could do alone. Think about the pyramids of Egypt. Think about the New York Philharmonic. Think about the fact that we put a man on the moon.

As these points make clear, our tendency to form groups beyond kin lines colors all facets of what it means to be human. Without this tendency for humans to form substantial bonds with others that extend beyond kin lines, we would be just another smart ape at the zoo.

How Throwing Rocks Paved the Way for Democracy

By definition, each species is unique. But let's face it: humans are particularly unique. A chimpanzee can shove a stick into a log and lure hundreds of termites onto the stick for a tasty, protein-laden treat. A person from New York can fly through the sky, land halfway across the world (in a matter of hours), get off the plane, and then immediately talk to his or her spouse on a cell phone. Humans are pretty unique. And, as you'll see, the origins of our uniqueness relates strongly to democracy in human governance systems.

So how did we get this way?

In an exemplary analysis of human origins, Paul Bingham and Joanne Souza (2009) made a strong case for human uniqueness as ultimately being the result of our ability to accurately throw rocks. This provocative thesis, summarized in detail in material that follows, takes data from an impressive set of academic areas into account. Their work integrated data from such fields as physical anthropology, evolutionary biology, evolutionary psychology, political science, and history—leading to some very provocative implications on the human condition.

Bingham and Souza made a strong case for humans as the democratic ape. To best understand our political nature, Bingham and Souza argued, we need to think of humans as being able to project a significant threat to others in a real and coordinated fashion. The authors called this the principle of *coercive threat,* and they argued that our ability to pose coercive threat to other humans is unlike anything that's ever existed in any other species. In short, they argued that humans evolved to be accurate and deadly in their throwing ability. We can throw rocks with much more deliberation, speed, and accuracy than can any other animal—by far.

Sound simple? Maybe irrelevant? Or even silly? Perhaps. But think about this: Imagine an ancestral group of hominids with a powerful but unfair and selfish leader who happens also to be bigger and stronger than any of the others in the group (which is how he got this leadership position in the first place). Attacking him physically would be risky; he can punch harder than you can—remember, he is big and strong. But throwing a rock, well that can smart a bit, and it doesn't have the same potential costs as close-up physical confrontation. In fact, thrown just right, a rock can kill someone. But the cost to the thrower is, again, small in terms of time and energy.

When humans first evolved the ability to accurately throw projectiles in this way, they gained the ability to hold coercive threat over others in an evolutionarily unprecedented manner. Humans could threaten conspecifics (i.e., other humans) from a distance.

Now couple this fact with another aspect of humans that is clearly part of our evolutionary story: the forming of social alliances. Humans are clearly social beings, and we often form alliances with others beyond kin lines. So now imagine a group of three or four males who are each smaller than the leader but who form a group, with a vision of creating a clan (or society) that affords them and their families more in the way of power and resources. This small group can be powerful. With shared vision and the ability to accurately throw from a distance, a small group can, in fact, be more powerful than a single large leader.

We often talk about there being strength in numbers. This is a foundational feature of what it means to be human, in fact. Teenagers look to their number of Instagram followers to assess their localized status. Candidates for elected office carefully examine data from polls to help them strategize. People often form coalitions in their efforts to effect political change at all levels. In *Homo sapiens*, there truly is strength in numbers, and Bingham and Souza's research tells us why that is.

The ability to emit coercive threat from a distance coupled with the proclivity to form significant social alliances may well have given rise to the nature of basic governance systems in human groups. In such a scenario, the social playing field can be leveled, and egalitarianism and democracy can emerge. Our natural tendency, then, may well be democracy.

Governance Systems Are Exploitable

While humans, then, may have a tendency toward working together in a coordinated fashion, history tells us that things don't always work out like peaches and cream. Human evolutionary psychology is a funny thing. Humans have clearly evolved a suite of features designed to help facilitate cooperation among people, such as our tendency to build non-kin–based groups of various kinds. This said, humans, like all organisms, have also evolved to have a suite of features that promote selfish interests, such as the tendency eat when hungry or to dodge an object that is coming at you, among many others.

So in examining human social systems and processes, we need to always keep in mind this dual-process nature of the evolution of human social behavior. Humans evolved to take steps to facilitate their own survival and reproductive success, *and* humans evolved to work together with others toward common goals.

Often, these competing features of our evolved psychology come into conflict with one another. Imagine you are a 5-year-old at a birthday party. You just loved the chocolate cake that was being served. Your friend Joey didn't get any cake, and you see that there is one piece left. You are starving, and there are no grownups in the room. What might you do? Remember, you are 5! Sure, you might eat the cake.

This example with the birthday cake can help us understand much about the evolution of human sociality. On one hand, 5-year-old you might eat the cake and make Joey sad. Or maybe, 5-year-old you has a well-developed sense of working toward the greater good and hands Joey the last piece of cake. Either is possible. The point is that this kind of decision permeates human life, and this is an important feature of our unique evolutionary heritage.

So now we can think about how this balancing of self versus group interests plays out in the modern political sphere. In a democracy, most elected officials usually run on some platform that underscores the greater good. A typical political catchline might be something like, "A vote for Smith is a vote for Springfield!" That would likely be much better than, "A vote for Smith is a vote for Smith's beach house!"

This said, as all humans evolved largely to take steps to benefit themselves, democratic systems are always exploitable. If you don't believe this, just think about the following names: Richard Nixon, Eliot Spitzer, Anthony Weiner, Recep Erdoğan, Idi Amin, Adolf Hitler, etc. Leaders who start out saying that they are ultimately going to take steps *for the people* don't always work out so well. The temptation to exploit leadership positions for one's own gain is ubiquitous. And given the fact that we evolved *both* to benefit ourselves and to benefit our broader social communities, human government systems sometimes are marked by democracy and sometimes by totalitarianism.

Our tendency toward democracy is exploitable and always runs up against the tendency for people to seek power and resources for themselves, which can lead to totalitarian governments that end up being a bad deal for the lion's share of the citizens overseen by that government. For this reason, members of a democracy need to stay vigilant. Any democracy can, under certain conditions, turn into a totalitarian government—history tells us this clearly. So while we may have natural proclivities toward democracy based on our uniquely groupish nature, democracies only work when all citizens take an active role.

Minds Evolved to Think Locally

One of the problems associated with modern politics is the fact that human behavior did not evolve with large-scale politics in mind. To begin to understand this concept, think about the following:

- When we hear that a nation halfway across the globe has adopted laws to reduce freedom of expression, we might give a shake the head.
- When we find that a state in our own nation is trying to limit some kind of rights based on sexual orientation, we might get mad.
- When a political problem erupts within our family system or at work, we might lose sleep.

In many ways, political situations get worse as they strike closer to home.

As we've discussed throughout this book, one of the core ideas that underlies evolutionary psychology is *evolutionary mismatch* (see Geher, 2014). This is basically the idea that the human mind evolved for generations under conditions that were, in many ways, very different from modern contexts. And when humans run into situations that are very discrepant from our *environment of evolutionary adaptedness* (i.e., the EEA), we are often not well equipped to deal with them.

Modern Politics Conceptualized in Terms of Evolutionary Mismatch

Per the concept of Dunbar's number, which we briefly touched on in Chapter 3, evolutionary psychologist Robin Dunbar (1992) provided evidence that our neocortex evolved to be able to accommodate information on about 150 people at a time. This fact makes sense as ancestral human groups that were nomadic in nature rarely exceeded 150. A straightforward implication of this finding is this: human minds did not evolve to deal with the large-scale social worlds of today.

In a study published in the *European Journal of Social Psychology* (Geher et al., 2015), summarizing work by our research team (the State University of New York New Paltz Evolutionary Psychology Lab), data are presented on how well people are able to cognitively process information related to large-scale versus small-scale politics. In this study, 49 college students read definitions and examples of large- and small-scale politics. Large-scale politics were defined as political situations dealing with states, nations, or international conflict, while small-scale politics were defined as the kinds of politics that our ancient ancestors might have run into, such as conflict within one's family or in a localized social group.

Participants were asked to write paragraphs describing four different kinds of political situations:

- situations that were large scale and not self-relevant
- situations that were large scale and self-relevant
- situations that were small scale and not self-relevant
- situations that were small scale and self-relevant

Our research team used a writing-sample analyzer to assess the nature of the writing used for these different kinds of tasks. The writing-sample analyzer provided quantitative information on such characteristics as the number of words used, number of sentences used, readability, and the grade level (e.g., sixth-grade level) that the writing was most suited for.

As we document in detail in the Results section of the article (Geher et al., 2016), we found that writing samples designed for large-scale political situations had more sentences and were less readable than were those designed for small-scale situations, and writing samples designed for small-scale (especially self-relevant) situations were written with the most fluidity. People seemed to have a much easier time writing and, thus, thinking about small-scale political situations compared with large-scale political situations. Our interpretation of the data is straightforward: The human mind did not evolve to take large-scale political situations into account as there were no large-scale politics that even existed for the lion's share of human evolutionary history.

Evolved for Localized Politics.   The human mind evolved for conditions that are, in many ways, very different from modern conditions. The modern world has such evolutionarily unnatural features as Taco Bell, Cosmopolitan magazine, cocaine, and socially defined groups that number into the millions. Modern politics often revolves around such large-scale social groups. One reason that modern politics often leads to epic failure is the fact that under ancestral conditions, political situations never addressed issues that pertained to more than 150 or so at a time.

As discussed throughout this book, the evolutionary psychological perspective has proven to have significant implications for such fields as education (see Gray, 2011); physical health (see Kruger & Nesse, 2006); mental health (see Glass, 2012); and other facets of life that underlie the human condition. Based on the research on the evolutionary psychology of politics described in this chapter, it is clear that the evolutionary perspective has much in the way of insights to provide in our efforts to better understand human politics.

An Evolutionary Perspective on Cultivating Good Government

Positive psychology has a history of focusing on what's good about the human condition. It is easy to criticize the government. As citizens who have a vested stake in our shared future, we would be wise to consider our evolutionary history in thinking about how we can, as the people, help improve governance systems to ultimately make things better.

In this chapter, three basic themes regarding the evolutionary psychology of politics emerged:

1. Humans have evolved a combination of self-interested and group-interested features, and this dynamic plays out in human political systems writ large.
2. Humans evolved the ability (through accurate throwing of projectiles) to form coalitions that have the capacity to keep selfish leaders in check and to give power to the people.
3. Human psychology is much better prepared to deal with small-scale rather than large-scale (modern) politics.

Each of these themes has significant implications for how people can positively affect modern politics.

First, based on the fact that humans evolved a combination of self-interested and group-interested features, we need to be wary of all elected officials. While skepticism is not always the most positive way to deal with people, when it comes to the people who represent our interests in government at all levels, the stakes are too high. Elected officials need to be vetted intensively during their campaigns and during their tenure in their elected positions. The tendency for people to exploit power for their own selfish interests, thereby not working for the best interest of the people, is simply too deeply rooted in our evolution to discard. While it would be a major mischaracterization to say that humans evolved to be all-out selfish, it would be naïve to think that all elected officials have the interests of the people exclusively in mind.

Second, humans evolved to embrace that fact that there truly is power in numbers. Humans evolved the capacity to form coalitions and to work together to influence change that benefits the broader group. This facet of human evolutionary history cannot be overemphasized. If you live in a democracy and you don't like the government, you have choices. Forming coalitions and working in a coordinated fashion to influence political situations is an ancient human story that runs deep into our evolutionary history. And this fact goes quite nicely with the First Amendment of the Constitution, which affords US citizens the rights to free speech, freedom of the press, peaceable assembly, and criticism of government. It also fits nicely with the history of labor unions, which capitalize on the concept of strength in numbers.

The third theme pertains to evolutionary mismatch and our inability to understand modern large-scale politics as well as we understand small-scale politics that have more ancient roots. This fact has important implications for so much when it comes to modern politics. For one, it can help us understand the famously low levels of voter turnout that we see in places like the United States (with millions and millions of registered voters choosing to stay home during the 2016 general election, i.e.). People often don't see how their particular vote matters; they don't see clear connections between what is going on in Washington, D.C., and what is going on in their particular lives; or the issues at hand are too complex for them to think about.

If we want to build nations of engaged citizens who understand the issues and who take steps to do their part in the process, we need to work with the fact that people don't naturally connect with large-scale political issues, in spite of how important or relevant they may be. Efforts, then, to get people involved in the political process need to regularly underscore how issues are relevant to the everyday lives of the people.

Working Toward the Greater Good

Ultimately, connecting positive psychology with issues of politics is all about focusing on ways that we can advance the greater good. That is, we need to think about how we can strengthen our communities for all citizens.

To put a face to this concept of the greater good and to think about it in terms of evolutionary principles, it is helpful to think about the famous Bostonian *tragedy of the commons* through a Darwinian lens (cf. Rankin, Bargum, & Kokko, 2007). The Boston Commons is a large park in the middle of Boston. Years ago, it was the main pasture area for cattle in the city. Grass became scarce as the city grew, and the Commons (meaning the common area) became the designated area for raising farm animals.

The land was owned communally (thus the term *commons*), and farmers owned their particular livestock. Well if you had three cattle grazing but your next-door neighbor owned four, you were in trouble. Your neighbor was going to sell more milk and meat than you would. So you might get a fourth head of cattle. Then the neighbor might get a fifth, then you might get a fifth, and so on.

Now imagine this scenario happening in all corners of the Boston Commons—over a few short years. Yup, your imagination is correct! The land cannot handle all of the livestock, and eventually it all becomes unsuitable for grazing. Pretty quickly, things fall apart, and everyone is fully out of business.

Thinking about the greater good—what is in the interest of the broader community over one's own particular interests—is not always easy. The proclivity toward focusing on one's own particular interests is often just too tempting. In a sense, it is this issue that underlies human politics at all levels.

Humans evolved with a combination of attributes that facilitate selfish interests as well as attributes that facilitate steps to help with the greater good. An understanding of evolutionary psychology can help us best build social structures that function for the greater good taking human nature into account.

Implications for Positive Psychology

The work reviewed here, addressing the evolutionary psychology of human politics, has many clear implications for the field of positive psychology. First, the work on the evolutionary origins of human coalitions has important implications for building positive and healthy communities. If we want healthy communities, we need to understand the evolutionary origins of human communities in the first place. This work also sheds light on the origins and nature of democratic decision-making—addressing how we evolved such that power is often distributed across all members of a community—and how achieving such an egalitarian state is a foundational aspect of healthy human communities. Finally, this work addresses the importance of evolutionary mismatch when it comes to human politics. To the extent that political outcomes affect our psychological lives, understanding the nature of evolutionary mismatch in modern politics is critical to understanding how to achieve positive psychological and communal outcomes.

Bottom Line

Politics—if you want a word that is charged in all kinds of ways, this is a good candidate. People get fired up about all kinds of politics: large-scale politics (Can you believe that the UN is not stepping in?); midscale politics (Is the state really going to cap local school budgets—is that even legal?!); small-scale politics (There is no way I'm going to vote for that guy for school board!); to very small-scale politics (Sally obviously didn't get the manager position over Ted because Ted is the owner's nephew!). Politics!

Work on the evolutionary psychology of politics has shed important light on the factors that truly underlie human politics at all levels. If we want people to be positively engaged in political processes at all levels and work toward the greater good in their approaches to life, then we simply need to take a Darwinian approach into account.

Acknowledgments

Some content from this chapter was adapted from Glenn Geher's (2014) *Psychology Today* blog post, "The Evolutionary Roots of Democracy." Other material is adapted from his (2015) post, "The Evolutionary Psychology of Politics." Glenn owns the copyright to the material.

References

Bingham, P. M., & Souza, J. (2009). *Death from a distance and the birth of a humane universe.* Lexington, KY: BookSurge.

Dunbar, R. I. M. (1992). Neocortex size as a constraint on group size in primates. *Journal of Human Evolution, 22,* 469–493.

Geher, G. (2014). *Evolutionary psychology 101.* New York: Springer.

Geher, G. (2014). The evolutionary roots of democracy. *Psychology Today* blog.

Geher, G. (2015). The evolutionary psychology of politics. *Psychology Today* blog.

Geher, G., Carmen, R., Guitar, A., Gangemi, B., Sancak Aydin, G., & Shimkus, A. (2016). The evolutionary psychology of small-scale versus large-scale politics: Ancestral conditions did not include large-scale politics. *European Journal of Social Psychology*, *46*(3), 369–376. doi:10.1002/ejsp.2158

Geher, G., Holler, R., Chapleau, D., Fell, J., Gangemi, B., Gleason, M., . . . Tauber, B. (2017). Using personal genome technology and psychometrics to study the personality of the Neanderthals. *Human Ethology Bulletin, 3*, 34–46.

Glass, D. J. (2012). Evolutionary clinical psychology, broadly construed: Perspectives on obsessive-compulsive disorder. *Journal of Social, Evolutionary, and Cultural Psychology, 6*(3), 292–307.

Gray, P. (2011). Free to learn. New York: Basic Books.

Kruger, D. J., & Nesse, R. M. (2006). An evolutionary life history understanding of sex differences in human mortality rates. *Human Nature, 74*, 74–97.

Rankin, D. J., Bargum, K., & Kokko, H. (2007). The tragedy of the commons in evolutionary biology. *Trends in Ecology and Evolution, 22*, 643–665.

Wilson, D. S. (2007). *Evolution for everyone: How Darwin's theory can change the way we think about our lives.* New York: Delacorte Press.

5

The Bright Side of Religion

Evolution and religion are famously antagonistic. The standard version of this conflict is about like the following: Most religions have stories about human origins built into their foundation. Christianity, Judaism, and Islam all go back to Adam and Eve in Genesis, for instance. The story there is not consistent—at all—with the perspective from evolutionary science. Evolutionary science, using data from the fossil record, DNA fingerprinting, and various other sources, suggests that the origins of life, and of humans ourselves, are the result of selection processes for organic materials that had the capacity to effectively replicate across excessively long periods of time. This is different from the idea that God created Earth and all its inhabitants in 7 days.

But as our friend David Sloan Wilson underscored in his work on the evolution of religion, the famous conflicts between scientists and religious fundamentalists are not particularly productive or necessary (see Wilson, 2002).

When we tell people that we conduct work in the field of evolutionary psychology, a common response is similar to the following: "Oh, I bet you get a lot of resistance from religious people!" And folks are often surprised when we tell them that the answer is "not really." There is plenty of resistance to the idea of human behavior resulting from evolutionary forces, but in our experience, the resistance is more likely to come from academics in other fields rather than from religious individuals (see Geher, 2006).

Don't get us wrong—there are definitely some religious people who hear the word *evolution* within the term *evolutionary psychology* and suddenly believe there to be some kind of conflict. Based on our years of experience in the field, we can confidently say that it doesn't have to be that way.

Where Evolutionary Psychology and Religion Collide

A few years ago, I (G. G.) agreed to take part in a debate with a Christian creationist about evolutionary psychology. My approach throughout was pretty simple: keep underscoring the fact that evolutionary psychology has nothing to do with creationism. Rather, evolutionary psychology is a powerful intellectual framework that can be used to advance the human condition along a multitude of fronts.

In thinking about it, it occurred to me that the work of evolutionary psychology really barely touches issues connected with creationism at all. And the primary point of conflict between religious individuals and evolutionists is right on this point—the point of conflict deals with issues of creation—the origins of life.

While evolutionary psychologists all take an organic, evolutionist approach to the origins of life on the one hand; on the other hand, we're really hardly concerned with this issue at all! Evolutionary psychologists spend very little time examining issues of the creation of life. It's really just not the focus of our field. Rather, evolutionary psychologists use a variety of evolution-based concepts, such as adaptationism, sexual selection, etc., to help us better understand the causes of human behavior. We can't think of a single evolutionary psychologist whose research focuses on the creation of life. This is one for the evolutionary biologists and paleontologists. Put simply: Evolutionary psychologists have lots of problems, but this ain't one!

The Effect of Religion on Endorsing Findings From Evolutionary Psychology

An interesting case in point regarding the interface of religion and evolutionary psychology is found in the work of Schwartz, Ward, and Wallaert (2011). These researchers presented a battery of findings from evolutionary psychology to two groups of adults. One group comprised religious individuals; the other group comprised secular individuals. The researchers asked the participants how much they believed in the various findings. In one condition, participants were simply given the findings. In another condition, participants were told that the findings *came from evolutionary psychology*.

The findings were quite revealing. In the *no-label* condition (i.e., when just presented with the findings), the religious participants were actually *more likely* to endorse the findings as accurate. However, get this: When the *ev-olutionary psychology* label was added, the pattern reversed, and religious individuals became *less likely* to endorse the findings. The data suggest that, in fact, religious individuals don't inherently have a problem with evolutionary psychology. Rather, they have a problem with anything with the term *evolution* in the title. This probably has to do with concerns regarding the conflict between evolutionist and creationist perspectives on the origins of life (which, again, are not relevant to the work of evolutionary psychologists).

The Evolutionary Psychology of Religion

In one of the most conciliatory scientific trends that we've seen in the past several decades, the field of *evolutionary religious studies* has emerged (see Wilson, 2002). The goal of this area of inquiry is hardly to discount or dismiss the beliefs and values of religious individuals. Rather, the goal of this field is to use the powerful framework of evolution to help us best understand a foundational aspect of humanity: the nature of religion. From the perspective of David Sloan Wilson (and several others in this field), religious beliefs and practices are foundational aspects of our species. Religions are found across the world, and they function similarly across cultures and across history. Religious practices have provided a basic means for humans to create social groups that extend beyond kin lines in a way that does not exist in any other primate.

Darwin's ideas on evolution are powerful, providing insights into so many areas of life (see Wilson, Geher, & Waldo, 2009). Religion is a basic feature of being human—and using Darwin's toolbox to elucidate the nature of religion has led to great insights into what religion is all about (see Wilson, 2002).

Scientific Fields and Religious Worldviews Differ in Their Goals

In thinking about the interface between evolutionary psychology and religion, it's important to keep in mind the very different goals of each of these entities. Evolutionary psychology is a scientific field of inquiry. The goal, as such, is to help advance our understanding of human behavior by applying evolutionary principles.

A religious doctrine has an entirely different goal, which is to provide a set of guiding principles on how people *should* carry out their lives. Without a set of moral principles (which may come in a number of forms, including both religious and secular), humans don't get along quite as well. Religious doctrines happen to be a primary mechanism for providing such guidance in our species.

A scientific field (such as evolutionary psychology) is all about explaining *how things are*. A religious doctrine is about providing guidance for how people *should* behave. These are, of course, very different kinds of goals, and understanding this distinction can potentially reduce perceived conflicts between religion and evolutionary psychology.

So, yes, evolution and religion are famously antagonistic, but they need not be. And evolutionary psychology in particular truly does not conflict with religion in any way as far as we can tell. In their work, evolutionary psychologists do not focus on issues of creation, and this is pretty much true across the entire field of evolutionary studies. This said, via the emerging field of evolutionary religious studies, evolutionary psychology actually has the capacity to shed much light on the nature of religion itself.

The Evolution of Big Deities and Religious Humility

One notable trend in religious traditions across time has been a move toward monotheism (see Shariff & Norenzayan, 2007). Monotheistic religions, such as Christianity, might be thought of as having a large and all-powerful deity. Shariff and Norenzayan's research focused on the evolutionary function of monotheism. In short, they explored the question of what adaptive benefits led to monotheism during a particular time in history.

Shariff and Norenzayan's 2007 analysis (based on the article, "God Is Watching You") is powerful and helps shed light on the nature of religion. These researchers argued the following: After agriculture became dominant, people didn't have to be hunters and gatherers; they could stick around where the food is. And once that happened, everything changed in a relatively short amount of time. In an evolutionary blink of an eye, people no longer had to follow the food, and people were starting to live in larger scale societies of thousands or even tens of thousands of individuals. Well, this environment that emerged was evolutionarily unprecedented in the species, and some things had to change as a result.

If your society now had to control 10,000 people in a shot, as opposed to 150 people, rules and other things needed to change. One thing that emerged during this time period was the single, powerful God of the primary monotheistic religions.

Why would a single, powerful, monotheistic God make more sense in a large-scale than in a small-scale society? Shariff and Norenzayan provided several examples to address this question. Perhaps the most vivid example pertains to the library at the University of Salamanca in Spain, a truly old university going back nearly 1,000 years. In a reading room that is highly remote from the rest of the library, evidence of students urinating on the walls goes back for centuries. As Shariff and Norenzayan put it, if you're deep in our studies and the closest outhouse is 10 minutes away, you may as well just pee on the wall.

But this outcome, of course, is not so great from the university's perspective. So what did the university administration do? They were smart. They started painting and sculpting saints all over the affected parts of the building. Suppose you're a religious, God-fearing student. Are you going to go ahead and pee on a saint? No. No you're not. That's because Christianity builds in enormous respect for the single God and his saints and disciples who comprise the all-powerful and all-knowing and who are responsible for everything. Don't mess with that. *Pay your respects, be humble, hold it in, and go outside!* And that is exactly what people did. Making God salient in this situation clearly facilitated prosocial behavior.

In large-scale societies, when controlling the behavior of a large number of individuals is foundational to success of the group (and the individuals therein), a single powerful and vengeful deity who elicits both respect and fear is going to have the capacity to cultivate prosocial behavior. Such an outcome will help the leaders of the group run the group without too much mutiny, and the ensuing organizational outcomes should feed back and, to some extent, benefit the individuals of a group that has got this whole *big deity* thing down.

The Grinch Who Stole Christmas and the Ultimate Function of Religion

Dr. Seuss had many a great insight into the nature of being human. If you want a poignant set of insights into the evolutionary function of religion, you need look no further than *How the Grinch Stole Christmas* (Seuss, 1957). A grouchy, nasty, selfish fellow, the Grinch concocts a plan to steal all of the presents from all of the kids of Whoville before Christmas morning. And, worse, he plans to take them for himself! This caricature of greediness and selfishness embodied in the Grinch speaks to several human archetypes. He's mean, rude, and without compassion. He's not in it for the greater good, by any means.

So of course the rest of the story is pretty much etched in the fabric of our culture by this point. The Grinch serendipitously bumps into Cindy Lou Who ("who was no more than two") when she genuinely mistakes him for Santa Claus, a much more generous soul than he. The Grinch seems to feel some pangs of guilt; he tells her something of a white lie about why he is shoving all of the toys up the chimney, and then he takes off, in true Grinch-like fashion.

Christmas, famously, comes in any case right on Christmas morning as scheduled, and all of the Whos celebrate gleefully and with appreciation, in spite of the lack of presents. The Grinch's heart famously grew by three sizes on learning of this news and seeing this situation unfold, and then he decides to turn to the bright side. He gives back all the toys, makes good with all of the Whos, and then, iconically, "he himself, the Grinch, carved the roast beast."

So there you have it: the Grinch in a nutshell. Now let's see how this American classic fits with an evolutionary take on why religion exists.

In his intellectually wrenching tour de force *Darwin's Cathedral*, David Sloan Wilson (2002) illuminated the basic functions of religion from an evolutionary perspective. According to David, religion, which characterizes any and all human societies that have been studied, has two important dimensions. The *vertical* dimension addresses the supernatural: *the what's up there* dimension of religion. This is the stuff that religious folks are asked to accept based on faith—supernatural deities, splitting of seas, reincarnation, and the like. Sometimes, the content of this dimension may seem strange to frame as possible: In fact, almost by definition, elements of the vertical dimension of any religion are impossible to reconcile with observable data. So here you have the *Hellfire exists because we as Christians believe it—it's in our books and we learn about it at church*—kind of thing.

For David, the vertical dimension of religion pertains to the *proximate* causes of religion, in other words, the immediate factors that make people religious. These include things like belief in a positive afterlife, belief in an all-powerful and good god, etc. These are things that encourage people to *do the right thing*.

However, for David, the ultimate function of religion pertained to social control. He called this the *horizontal dimension* of religion (as it extends across people). Religions have extensive codes about human behavior and social relations: *respect your mother, don't kill, don't trespass on your neighbors, be kind to strangers, treat others as you would treat yourself or your own family, share your materials with others in need*, etc.

A core feature that underlies all of these rules is essentially this: Subordinate your selfish interests for the greater good. If you can create a religion that gets people to act in the interest of others—and, ultimately, in the interest of the greater good—you will have a population of altruists with few folks showing all-out selfish tendencies. And, of course, as a group, this is exactly what you need to thrive. So religion's ultimate function, then, can be conceptualized as getting people to work toward the interests of the broader group (via social control, or horizontal mechanisms), partly by getting them to believe that some all-powerful deity is keeping an eye on their actions (i.e., the vertical dimension of religion).

Interestingly, this evolution-based approach to religion connects well with the work of positive psychologists Graham and Haidt (2010), who acknowledged the evolutionary origins of religion and morality and who argued that religious practices ultimately serve to unite people in forming communities based on shared moral principles.

So let's go back to our friend the Grinch. In short, the Grinch starts out as selfish and ends up as other oriented, and we as audience members like this shift. We are humans, and we like others whom we can expect to be working on our behalf and on behalf of the greater good. We often find those who are only working in their own self-interests to be repulsive. These facts about social perception connect closely with Wilson's take on the evolutionary psychology of religion. What sparked the Grinch to have a change of heart? It was the coming of Christmas, which is, for Christians, of course, a pretty important religious holiday—signifying the birth of Christ, a prototype of the vertical dimension of religion.

Seuss actually delved into what evolutionists would refer to as both *proximate* and *ultimate* mechanisms that relate to the Grinch's famous turnaround. Remember, the Grinch wasn't a nice guy until "his heart grew three sizes." Think about that! In this case, as is often used in our lexicon, a *big heart* is used metaphorically to represent a compassionate, other-oriented mind, or overall pattern of behavior. And this is exactly the kind of proximate mechanism that religions incorporate to facilitate the ultimate outcomes of actual other-oriented behavior.

Anthropological research on the evolutionary origins of religion often uses the sharing of food as a standard exemplar of other-oriented behavior. And how does the story of the Grinch end? That's right: He carves the roast beast himself. There you have it.

Religious Experience and Positive Affect

A core goal of much work in the field of positive psychology is to help culti-vate conditions that facilitate positive affect. In fact, research into the proxi-mate goals of religious experiences has shown that many aspects of religion pertain to exactly this process.

In a landmark study of the psychophysiology of the Pentecostal church experience, Chris Lynn, Paris, Frye, and Schell (2010) studied 52 Apostolic Pentecostals from a rural area of New York. All of these participants had ex-perienced the *speaking-in-tongues* phenomenon during church services. If you're not familiar with this phenomenon, you might want to check it out on YouTube. It is essentially a phenomenon marked by someone getting so wrapped up in the religious experience during a church service that he or she gets up and starts speaking loudly and passionately—often in an indecipher-able manner—as if he or she is channeling some facet of God himself.

The main question of Lynn et al. (2010) was this: Why do these people do this? Or, from a more formal, scientific perspective, they asked if there is any adaptive, functional benefit to such behavior. To address this question, they actually spent many hours attending Pentecostal services, making a point to gain the trust of the parishioners before studying them directly. Chris Lynn (we know him personally) is a supernice guy, so this wasn't too difficult for him. Once the research commenced, Chris moved forward and started with data collection. Via analysis of the parishioners' saliva, Chris studied two im-portant stress markers: salivary cortisol and α-amylase. Effects were found for α-amylase as follows: Levels were lower on worship days compared with on nonworship days. Chris concluded that highly impassioned, emotion-ally charged worship experiences served as having a proximate physiological benefit: reducing markers of stress. And as we have seen across this chapter, proximate outcomes are often tied to ultimate outcomes when it comes to evolutionary findings. So reductions in stress famously correspond to all kinds of physiological and psychological outcomes. In short, reductions in stress ultimately facilitate survival and reproduction, or Darwin's bottom line.

Stress reduction via intensive worship service is not the only way that religion provides emotional benefits to individuals. A great deal of research has shown that forgiving others, which is a core value across religious sects, feels good (see Gorsuch & Hao, 1993) and corresponds to a variety of psychological benefits, including a reduction in stress. Forgiveness also serves as a buffer between stress and health outcomes, so a forgiving person is more likely to weather a stressful storm in terms of his or her mental and physical health compared with his or her less-forgiving counterpart.

As we elaborate in Chapter 6, forgiveness is one of the basic moral emotions and is rooted deep in our evolved psychology. Religions across the globe underscore forgiveness as a basic aspect of any approach to dealing with others. From the perspective of a religious group, having forgiving parishioners keeps harmony within the group and, thus, keeps the group members strongly bonded to one another—poised to help advance the goals of the group. From an individual's perspective, forgiving keeps one connected with others in the group, which is a basic outcome associated with success in a highly social species such as ours. Through an emphasis on forgiveness, religions, then, cultivate various social processes that ultimately benefit oneself and one's broader community.

Religion as Psychology's Double-Edged Sword

As you can see by the content of this chapter, evolutionists such as we are take an amicable approach to religion. This said, we'd be naïve to think that religion is all peaches and cream. It is without question that religious ideology of various kinds has wreaked havoc on people's lives across the world for thousands of generations. While religion may have a bright side and be part of the foundation of human moral psychology, it's important to understand why, from a deep evolutionary perspective, religion so often leads to problems.

The tendency for modern *Homo sapiens* to form large social groups beyond kin lines bears strongly on the fact that humans are the churchgoing ape. And while there is much variability from one religion to another in terms of specific practices, all religions (see Wilson, 2002) seem to cultivate this uniquely human tendency to form psychologically created groups that extend beyond kin lines.

One of the main psychological antecedents to religious thinking is what we call *in-group/out-group reasoning*. People around the globe are highly predisposed to automatically put people into two basic groups: *my group* or *not my group* (see Billig & Tajfel, 1973). This mentality helps benefit religious groups (and, indirectly, the individuals within them) because it keeps members of the same group positively predisposed toward one another. Interestingly, our Neanderthal cousins likely did not have this same tendency to form psychologically created groups beyond kin lines that we often connect with religion. And this is often framed as one of the main reasons that we don't see many Neanderthals around these days (see Geher et al., 2017).

This said, when you look at the bowels of human history, you frequently find evidence that in-group/out-group reasoning is not always a bowl of cherries. This kind of reasoning seems to predispose members of one group to see members of other groups (out-groups) in less positive terms. This reasoning also corresponds to the tendency for people to see members of out-groups as sort of "all the same"; we call this tendency *out-group homogeneity*, and we often can't help it! It is hard for us to see members of groups that are very different from our own in the same positive ways that we often see members of our own groups. This all seems to be part of the basic social psychological architecture that evolved along with origins of religious belief in early *Homo sapiens*. In-group/out-group thinking and the tendency to see all out-group members as similar to one another are foundational features of human social psychology; like it or not, it's part of how we think.

In his famous book *Cat's Cradle*, Kurt Vonnegut (1963) explicated these concepts well. Particularly, he came up with the idea of a *granfalloon*, which essentially is an artificially created group based on some meaningless characteristic that binds people together psychologically. So in *Cat's Cradle*, for instance, when strangers on a plane trip to a tropical island find out that that they happen to both be from Indiana, they are suddenly bonded—extolling the virtues of being a Hoosier and progressing as if they now can trust one another for life. Vonnegut called it granfalloon, social psychologists call it *in-group/out-group reasoning*. It's the same thing; it's the way that we automatically form groups based on minimal criteria and become predisposed toward members of these groups no matter how silly these criteria are (sorry Hoosier nation!).

In-group/Out-group Reasoning: The Good and the Bad

Like many evolved features of human psychology, in-group/out-group reasoning comes with a price. Without the capacity to form large-scale, coordinated groups—thanks partly to in-group reasoning—we would not have the Eiffel Tower, the Internet, or Ben and Jerry's ice cream. All of these great wonders of the modern world were created by coordinated groups of humans working together toward a common goal—large groups of predominantly unrelated individuals who saw themselves as part of something bigger and as connected to one another. This is the primary benefit to our world for which we can thank our evolved in-group/out-group reasoning.

But in-group/out-group reasoning has a dark side, and the long history of war that goes thousands of generations deep in our species betrays this fact. As philosopher David Livingstone Smith (2008) showed, war is something of a human constant (and the tendency for warring individuals to devalue members of the other side, seeing such out-group members as vermin or as nonhuman in some other way) is remarkably common in the social psychology of warfare. Seeing your enemies as nonhuman makes it much easier to kill them, and we can thank our in-group/out-group reasoning for this as well. It is without question that in-group/out-group reasoning bears on the nature of religion in humans. Religion may have a bright side, but we must always be vigilant of the fact that religion can have a dark side as well.

Implications for Positive Psychology

Positive psychology, focusing on factors that are associated with human thriving at the individual and community levels, needs to understand the evolutionary origins of human religion to fully appreciate the origins of the positive aspects of the human experience. The following is a list of concepts from work on the evolutionary psychology of religion that positive psychologists would benefit from incorporating in their work:

- Religion is a basic part of the human experience, existing in thousands of forms across the globe.
- As an evolved set of psychological mechanisms, religious practices consistently have strong implications for personal growth as well as for the cultivating of a community.

- Religion has two basic axes: the vertical dimension, which functions to connect people with some supernatural phenomenon that is more powerful than they are, and the horizontal dimension, which functions to maintain social control over people.
- Monotheism seems to have emerged particularly to exert social control over large numbers of people once civilizations emerged in a postagricultural world.
- A core way to think about religion from an evolutionary perspective is to see any religion as having the ultimate goal of getting people to subordinate their selfish inclinations for the good of the group.
- Religious experiences have all kinds of positive psychological and physiological effects, a fact that speaks to the ultimate evolutionary benefits and origins of religion.
- Religion can lead to unhealthy levels of in-group/out-group psychology, a fact that underlies the dark side of religion in human societal functioning.
- Religion emerged largely to help people and communities thrive; thus, in many ways, religion evolved in humans in a way that parallels the goals of the positive psychology movement.

Bottom Line

For more than a century now, religion has famously been at odds with evolutionary science. However, under the leadership of such scholars as David Sloan Wilson, many evolutionary scholars are turning away from the idea that religion and evolutionary science are somehow irreconcilable or at odds with one another in any important way. This relatively new approach to evolutionary religious studies focuses on how we can use the insights of evolutionary science to better understand the origins and functions of religion.

From an evolutionary perspective, religion evolved to encourage individuals within groups to act in a selfless way, putting the goals of the group ahead of their own goals. By understanding this basic feature of human evolved psychology, positive psychologists will gain keen insights into the evolutionary roots of the phenomena that they are studying. If you want to understand the positive aspects of the human experience, understanding the evolutionary origins of human religion is essential.

Acknowledgments

Some content from this chapter was adapted from Glenn Geher's (2016) *Psychology Today* blog post "Evolutionary Psychology Goes Just Fine With Religion"; his (2014) *Psychology Today* blog posts "You're Going to Have to Serve Somebody"; "Broadway's Book of Mormon (More Offensive Than Promised)" (2013); and "Why Are There More *Homo sapiens* Than Neanderthals These Days?" (2015). Glenn owns the copyright to the material.

References

Billig, M., & Tajfel, H. (1973). Social categorization and similarity in intergroup behaviour. *European Journal of Social Psychology, 3*, 27–52.

Geher, G. (2006). Evolutionary psychology is not evil . . . and here's why. *Psihologijske Teme (Psychological Topics); Special Issue on Evolutionary Psychology, 15*, 181–202.

Geher, G. (2013). Broadway's Book of Mormon (More offensive than promised). *Psychology Today* blog post.

Geher, G. (2014). You're going to have to serve somebody. *Psychology Today* blog post.

Geher, G. (2015). Why are there more Homo sapiens than Neanderthals these days? *Psychology Today* blog post.

Geher, G. (2016). Evolutionary psychology goes just fine with religion. *Psychology Today* blog post.

Gorsuch, R. L., & Hao, J. Y. (1993). Forgiveness: An exploratory factor analysis and its relationship to religious variables. *Review of Religious Research, 34*, 351–363.

Graham, J., & Haidt, J. (2010). Beyond beliefs: Religions bind individuals into moral communities. *Personality and Social Psychology Review, 14*, 140–150.

Lynn, C. D., Paris, J., Frye, C. A., & Schell, L. M. (2010). Salivary alpha-amylase and cortisol among Pentecostals on a worship and nonworship day. *American Journal of Human Biology, 22*, 819–822.

Schwartz, B., Ward, A., & Wallaert, M. (2011). Who likes evolution: Dissociation of human evolution versus evolutionary psychology. *Journal of Social, Evolutionary, and Cultural Psychology, 5*, 122–130.

Seuss, T. (1957). *How the Grinch stole Christmas*. New York: Random House.

Shariff, A. F., & Norenzayan, A. (2007). God is watching you: Supernatural agent concepts increase prosocial behavior in an anonymous economic game. *Psychological Science, 18*, 803–809.

Smith, D. L. (2008). *The most dangerous animal*. New York: St. Martin's Griffin.

Tinbergen, N. (1963). On aims and methods of ethology. *Zeitschrift für Tierpsychologie, 20*, 410–433.

Vonnegut, K. (1963). *Cat's cradle*. New York: Dell.

Wilson, D. S. (2002). *Darwin's cathedral: Evolution, religion and the nature of society*. Chicago: University of Chicago Press.

Wilson, D. S., Geher, G., & Waldo, J. (2009). EvoS: Completing the evolutionary synthesis in higher education. *EvoS Journal: The Journal of the Evolutionary Studies Consortium, 1*, 3–10.

6

Happiness, Gratitude, and Love

Few things in the world are certain, but some things are. On the short list of things that are certain is the fact that life is fleeting. Given that we only have one chance here, we need to think hard about what goals we need to set for ourselves and work toward. Life goals vary across cultures and across people. A common goal in Westernized societies is the pursuit of happiness. In fact, the US Constitution underscores the pursuit of happiness as one of the foundational rights in any free society.

With a focus on the positive aspects of psychological life, positive psychologists have largely focused on factors associated with human happiness (see Watkins, 2014). On the surface, this makes perfect sense. Given the nature of the human emotion system, it's clear that humans have evolved to move toward stimuli that facilitate happiness and away from stimuli that facilitate such outcomes as anger or sadness. In one of the first landmark expositions of evolutionary psychology, Charles Darwin himself made this case (see Darwin, 1872).

In fact, a key to ultimately understanding the optimal psychological goals of life takes human evolution deeply into account. Only by integrating an evolutionary perspective into questions about life goals can one begin to understand (a) why the human emotion system evolved as it has and (b) how the evolutionary history of human emotions can shed light on what we should all be striving for during our short-lived time here.

As we'll soon see, the evolutionary perspective has extraordinary insights to shed on the nature of human happiness. Happiness is not bad, but from an evolutionary perspective, it is clearly a proximate goal rather than an ultimate evolutionary goal. And this point has enormous implications for a positive psychological approach to life.

Are We Evolved for Happiness?

In a 2017 presentation in the evolutionary studies seminar series at the State University of New York at New Paltz, renowned psychiatrist Randolph Nesse told a simple anecdote that speaks volumes about the nature of happiness. The story was essentially as follows:

> A while back, Nesse had a client who was a professor. This professor was, like so many of us, dealing with issues of anxiety. Nesse prescribed antianxiety medication. A few months later, Nesse asked the client how he was doing, and the professor said that he felt great for the first time in years. This said, he indicated that there was one problem: A huge stack of student papers had been sitting on his desk for weeks, but he had no motivation whatsoever to grade them.

In his landmark work on evolutionary medicine (Nesse & Williams, 1995), Nesse made the case that symptoms, be they physical or emotional, typically exist for good evolutionary reasons, and that any medical approaches that focus exclusively on ameliorating symptoms are naïve, limited, and potentially problematic.

In the case of the anxious professor, it seems that the anxiety he was experiencing, while unpleasant, actually played a motivational factor in his life: *It motivated him to do his job.*

In fact, from an evolutionary perspective (Nesse & Ellsworth, 2009), this is pretty much the ultimate reason that anxiety exists at all: It plays a factor in motivating people to take actions that are ultimately adaptive. Under ancestral conditions, people likely became anxious under the following kinds of conditions:

- They experienced snakes in the jungle.
- They found themselves lost in the woods.
- Others in their social circles shunned them.
- Their social status was threatened.
- They experienced breakups with intimate others.

Ancestors who showed no anxiety to these kinds of threats would not have taken appropriate steps to fix the problems. Someone who was not anxious at the thought of a poisonous snake attacking would be less likely to avoid these dangerous animals compared with anxious-prone others. Someone who felt no anxiety when his or her social status was publicly threatened would not be likely to take steps to restore his or her status within the group, and so on.

Anxiety is not pleasant, but it is a substantial feature of human psychology for a reason. And reducing anxiety fully, as demonstrated by the anecdote regarding Nesse, can actually be *maladaptive* and counterproductive.

Happiness Is Only a Proximate Goal

From an evolutionary perspective, human emotions evolved as they have because they generally work to confer evolutionary benefits to us, helping to ultimately increase the likelihood of survival or reproductive capacities (see Guitar, Glass, Geher, & Suvak, 2018). Just as anxiety evolved to help motivate adaptive behaviors, happiness has also evolved to help motivate adaptive behaviors.

If you look at the things that make people happy, you can quickly see that they generally map onto outcomes that would have, on average, led to increased probabilities of survival or reproduction for our ancestors. Here is a short list:

- Food
- Sex
- Engaging in fun times with friends
- Success in social contexts
- Task completion

In broad strokes, we can easily see how these kinds of outcomes not only have the capacity to lead to happiness, but also have clear benefits from an evolutionary perspective. The evolutionary take on happiness, then, is essentially this: Happiness is an affective state that motivates us to engage in actions that are likely to lead to outcomes that would, on average, lead to increases in the likelihood of survival or reproduction (see Guitar et al., 2018).

In evolutionary parlance, we would say that happiness is a *proximate outcome*. It matters and it is nice, but it is *not* an ultimate evolutionary outcome.

From an evolutionary perspective, ultimate outcomes pertain to outcomes that bear on increases in the likelihood of survival and reproduction. Thus, we may be thrilled to have a piece of chocolate cake on the table in front of us at a birthday party, but that momentary happiness is not an end goal in itself. We evolved to be happy when presented with rich food offerings because our ancestors, who were motivated to find rich foods, were more likely to eat and thus to survive and reproduce.

Happiness, then, like anxiety, is an affective state with the primary evolved function of motivating us to engage in behaviors that would have led to evolutionarily adaptive outcomes under ancestral conditions. Happiness is not an end goal; it is *a means to an end.*

Naïve Positive Psychology Versus Positive Evolutionary Psychology

In recent decades, there has been much growth in the area of positive psychology (McMahan et al., 2016). Positive psychology largely focuses on helping us better understand the positive aspects of the human experience. Without question, we support this broad mission of positive psychology and are appreciative of the luminaries like Martin Seligman and Scott Barry Kaufman who are doing great work to help advance this area of inquiry.

This said, once you think about the human psychological world from an evolutionary perspective, it can be seen that positive psychologists need to be cautious about overemphasizing factors that increase happiness. From this perspective, work that focuses on increasing happiness without a well-grounded *evolutionary* perspective misses the boat. For a psychology of happiness to truly advance, it must take evolutionary principles and research into account. Positive psychology that focuses on increasing happiness without taking evolutionary context into account is simply naïve and potentially misguided. Or, as Todd Kashdan and Robert Biswas-Diener (2014) put it, *there is an upside to your dark side.*

Happiness, Positive Psychology, and Darwin's Lens

This section on the evolutionary psychology of happiness is strategically placed near the center of this book. Positive psychologists have gained much ground in the past several decades, helping elucidate several basic features of human psychology that lead to thriving and success in life. This said, any large-scale analysis of the field of positive psychology turns up a major disconnect when it comes to the field of evolutionary psychology. This disconnect is, to understate the point, a major problem. Increasing happiness is not a bad goal by any stretch of the imagination. But it is not the *end* goal, and it's hardly the point of life. If we want a broader understanding of what it means to be human, we need to include an evolutionary perspective.

Living a Rich Social Life

As demonstrated in the previous material of this chapter, the evolutionary perspective can be extremely helpful in understanding one's internal world: helping to understand the nature of such important emotional states as happiness. As discussed in the subsequent sections of this chapter, the evolutionary perspective also has much to say when it comes to living a rich and rewarding social life. When it comes to dealing with others and getting along in a broader social context, the evolutionary perspective has proven both powerful and poignant.

To understand the evolutionary perspective on living a rich and meaningful social life, let's go back to the 1970s. In 1976, renowned biologist, thinker, and writer Richard Dawkins published *The Selfish Gene*. The book is a highly accessible and powerful summary of Darwin's ideas on evolution, applied largely (but not fully) to several classes of animal behavior, including the mating habits of the praying mantis, the murderous nature of emperor penguins, and the helpful nature of vampire bats. *The Selfish Gene* was a game changer when it was published, providing the guts of Darwin's vision of life in a highly readable paperback. Thousands and thousands, if not millions, of people in the world today understand the basics of evolution thanks to this book.

One intellectual consequence of Dawkins's (1976) provocative title for this book, which has substantial implications for what it means to live a rich social life, was a focus on the many connotations of the term *selfish*. He meant the term in a very specific sense: A *selfish gene* is one that codes for qualities of an organism that increase the likelihood of its own survival or reproductive success. In short, in the future of a species, replicating genes out-exist nonreplicating (or poorly replicating) genes.

This is really all he meant.

As history tells us, Dawkins (1976) had many critics of this work. The reasoning of critics of *The Selfish Gene* argued that if genes that exist are selfish, then the products of those genes, including humans like you and me, must be selfish, too. This fallacious reasoning has driven much of how evolutionary science has progressed since the publication of Dawkins's book and how evolution (now seen by many as espousing a "red in tooth and claw" take on our kind) has taken on something of a cold angle on what it means to be any kind of organism, including a human.

The bad news is that this misinterpretation (or at least this overly applied extension) of Dawkins's metaphor, has not helped the publication relations of evolutionary science. People from the outside looking in often think, "Evolutionary science? Isn't that the stuff that says we are animals and we want to kill each other for our own selfish gain?"—not so pleasant a portrait.

But there is good news resulting from these trends as well, in the form of a landslide of research over the past several decades shedding light on the *positives* of human nature, and the book that you hold in your hands now is an exposition of this work.

The Reciprocating Ape

In 1971, renowned evolutionary behavioral scientist Robert Trivers published a groundbreaking theory of altruism that would forever change our understanding of prosocial behavior across species.

The most common form of helping found in the animal world—including the human world—is the helping of kin (Hamilton, 1964). In other words, people tend to help those with whom they share large proportions of genetic materials (e.g., mothers helping their infants). This makes good evolutionary sense as helping kin helps your genes as they exist in the bodies of others.

But kin-based altruism is only a slice of the helping behavior we see in humans (see Geher, 2014). As discussed in Chapter 4, addressing the great evolutionary leap that *Homo sapiens* has made, it is true that we often help non-kin. Some of us, such as social workers, teachers, and counselors, do this professionally—helping non-kin all day long for 5 days a week. All of us help non-kin with some level of regularity. Consider the following examples:

- Helping a neighbor move a piece of furniture.
- Holding the door for someone who's holding a bag of groceries.
- Letting a fellow driver enter your lane.
- Making a nice dish that you bring to a potluck meal at work to share with others.
- Sharing your notes from a lecture with a colleague who was sick during the last meeting.

We help non-kin all the time. As evolutionists, we can think of this fact in a few ways. First, to the extent that the helping of non-kin is species typical, then this feature of our species must be rooted in our ancestral past. Second, if the helping of non-kin is so central to our nature, then there must be something adaptive about this phenomenon. It must, somehow, end up paying back the helper in the end.

Humans are not the only primates that show non-kin altruism. In fact, there are many species that demonstrate this phenomenon. In the evolutionary literature, this phenomenon is often referred to as *reciprocal altruism*, a concept made famous by Trivers in 1971.

Put simply, reciprocal altruism is the helping of another with an implicit (unstated) agreement of having the help paid back. If you scratch my back, you can rest assured that I'll scratch yours when called on. People don't help all non-kin equally; we are discriminating. We are more likely to help non-kin who have helped us in the past. In using such an algorithm in helping non-kin, this form of altruistic behavior, then, becomes adaptive for the helper, and it is partly for this reason that helping of non-kin is so prevalent in our species.

Famous for having a knack for articulating basic evolutionary truths that characterize species of all shapes and sizes, Trivers demarcated three basic features that must be present in a species if reciprocal altruism is to be evolvable (i.e., capable of producing adaptive benefits to individuals) in that species:

1. **The members of the species must have relatively long life spans.** If the life span of the organism in question is terribly short, then helping another with the implicit expectation of receiving help in return is

not likely to pay off; your would-be altruistic friend is too likely to die young before being any use.

2. **The members of the species must be able to identify one another.** If you implicitly expect any kind of help in return for an altruistic act, then the individual you help had better be able to discriminate you from all the others! Not all species have this ability. And, as you can see, this ability is necessary for reciprocal altruism to be evolvable.

3. **The members of the species must live in relatively stable groups.** In some species, individuals are relatively solitary—perhaps except during mating season. If you are such a species, helping out a conspecific (i.e., a member of the same species) will not likely have much value to you. You'll never see that one again!

Do these three qualities typify our kind? You bet. And this fact helps us understand the highly prevalent nature of non-kin altruism in humans.

Are humans the only species that fit these three criteria? Not by a long shot. A short list of other species that fit these three criteria includes

- North American crows
- Guppies in the freshwater lakes of upstate New York
- Olive baboons
- Wolves
- Vampire bats

And in demonstrating Trivers' brilliance as an evolutionary biologist, guess what? Non-kin altruism exists in all of these species. Take vampire bats: The best-studied non-kin altruism in vampire bats takes the form of food sharing. Their mechanism of food sharing seems pretty gross to us, but it's really a great example: Not all vampire bats are successful in finding a blood meal each night. These unsuccessful bats, of course, get hungry. They are reliant on the successful hunters to go out, bite some livestock, get some blood, bring it back, and regurgitate in a helpful manner so that these less successful bats can get a meal. But the successful hunters don't share with just any old bat. They are biased. They are much more likely to offer up this generous regurgitation for specific others who have provided just this same generosity to them in the past.

While our table manners may be more civilized, humans are a lot like vampire bats: We show a bias toward helping others who have helped us in the past.

While there are many features of our social ecologies that serve as foundations for who we are, the reality of reciprocal altruism is perhaps one of the most significant. The evolutionary history of reciprocal altruism in humans clearly goes back deep into our history, and it helps explain so much about who we help, when we help, and when we decline to help. While humans are many things, with his ideas on reciprocal altruism, Trivers clearly made the case for understanding us as the "you scratch my back, I'll scratch yours" ape.

How Reciprocal Altruism Shaped the Human Social–Emotional World

In his classic textbook on the evolution of sociality, Trivers (1985) elaborated on how our long history of reciprocal altruism shaped our social and emotional worlds. The concepts developed therein have dramatic implications for understanding positive psychology from an evolutionary perspective.

Trivers essentially made the case that if you are a member of a species that stands as the hallmark of a reciprocally altruistic animal, then an entire suite of emotions and behavioral patterns will make sense as being selected to go along with this fact. We can think of these psychological processes and emotions as the moral emotions—emotional and social processes that serve the primary function of helping individuals maintain harmony and achieve status within their localized social circles.

Given the facts that humans (a) evolved primarily in small-scale social settings and (b) have a long history of within-species reciprocal altruism, it would have benefited ancestral humans to have psychological and behavioral patterns that mapped onto this reality. Ancestral humans who took an all-out selfish approach to social life would have done poorly in the long run. On the other hand, someone who developed a reputation as an altruist—as a self-sacrificer—would have actually benefited in a long-term manner.

In human social contexts, selfish and altruistic approaches have both costs and benefits. A selfish approach to living makes sure that one acquires short-term gains, but such an approach may lead to long-term costs, such as being repulsed by others or, in a worst-case scenario, being ostracized by the group—left to fend for oneself.

As for an altruistic approach, in the short term, an altruist does poorly. He or she is the last one in line, gets the last slice of pizza, and is the one who ends up missing the elevator, saying gently, "It's fine; I'll just wait for the next one." But being an altruist—and cultivating a reputation as an altruist—comes with great long-term benefits in a species such as ours.

Think about it this way: Imagine that there are two men who are totally equal in all ways except that one is prototypically selfish and the other is the quintessential altruist. Who is going to be picked to be captain of the team? Who do you want to be the manager of your group? Who do you want to marry and spend the rest of your life with? Who do you trust to watch your kids? Or, in a more raw format, who do you like better?

Human psychology has evolved so that we have a preference for altruists. This fact is a result of our long evolutionary history as reciprocal altruists; it colors so much of our social and emotional worlds.

Our complex emotions are largely rooted in our history of reciprocal altruism. Consider guilt. We are more likely to feel guilty when we have done something that is selfish compared to if we have done something that is altruistic. You are more likely to feel guilty if you have cheated on your spouse compared to if you have just folded her laundry, for instance. Guilt, then, is largely an emotional signal that tells us that we have done something selfish—that we have slighted someone in our social circle. It is a signal that makes us feel bad, and, like all emotions, it motivates behavior.

In this case, guilt motivates *reparative altruism*, behaviors that work to restore social bonds that we have, for one reason or another, weakened or damaged. Apologizing runs deep in humans. If you don't believe that, think about parenting. Think about how often you hear parents tell their kids that they need to apologize for something. This is, of course, a very regular occurrence, speaking to the fact that repairing relationships that need repairing is a basic part of human social life.

Men who forget their wedding anniversaries keep florists in business. In fact, we have entire parts of our lexicon that are all about reparative altruism. We say things like the following:

- "I owe you big time!"
- "You took a bullet for me."
- "I cannot thank you enough."
- "I am so sorry. I mean it."
- "You saved my life."
- "I will never ever do that again!"

On the flip side of guilt and apology we have forgiveness. We are encouraged to forgive others—so much so that forgiveness makes it into the fabric of major religions around the world. And just as apologetic behavior keeps us connected to others in our social worlds, forgiveness does the same. Forgiving someone signals that you will continue to accept that person in your social world in spite of some transgression.

Gratitude has a similar function: showing someone that you notice and appreciate his or her selfless actions. Showing gratitude toward someone can have the function of not only keeping one connected to others in the group, but also encouraging that other to continue to demonstrate selfless actions moving forward. Gracious people are those who show gratitude, and we evolved to favor gracious people.

Ultimately, gratitude can promote the formation and maintenance of positive and meaningful relationships (Algoe, Haidt, & Gable, 2008). Further, research that recognized gratitude as an adaptive evolutionary mechanism has also shown that gratitude serves to promote subjective well-being across the entire life span (Chopik, Newton, Ryan, Kashdan, & Aaron, 2017).

Apologetic, forgiving, and gracious behaviors serve similar functions. They keep people connected to others in their social circles. Someone who apologizes is signaling that he or she is really group oriented and not selfish. Someone who forgives signals the same, as does someone who demonstrates social grace. In a world of reciprocal altruists, these qualities are, of course, highly appreciated, and having a reputation as having these qualities helps individuals enormously in the long term.

Human social psychology is largely about taking steps to make sure that we remain in the good graces of others. Our particular social evolutionary history tells us why.

Humans as Loving, Helpful, and Self-Sacrificing

In light of this conception of humans as the reciprocally altruistic ape, here are just a few directions that evolutionary psychology has followed in painting the positive aspects of humans:

1. **Paying it back.** Giving back to others who have given to you in some important way is hugely significant from the perspective of evolutionary psychology. As discussed previously, Trivers's (1971) landmark

work on the topic of reciprocal altruism demonstrated that in relatively long-lived species, such as our own, the tendency for altruism among non-kin may evolve, and it may take the form of people helping others, even strangers. Sometimes this kind of help is *paying it back*, or reciprocating altruistic acts that have come to would-be altruists in a small social community. Not paying back altruism is socially dangerous— in your social ecosystem and in the social ecosystems of pre-agrarian humans all around the globe. *We've evolved to pay it back.*

2. **Paying it forward.** This is a term that's been thrown around a lot in recent years, and we love it! It essentially says to give to others—not to reciprocate them for having helped you in the past, but to help them proactively so that they are on good footing moving forward. Maybe they will help you in the future. Maybe they will help others close to you in the future. Maybe they will help the broader community. Your helping them proactively sets the stage for any of these outcomes with the potential to positively influence you, your kin, and your social network. Paying it forward is seen positively in social communities; it helps people develop reputations as altruists or helpers or, more simply, folks whom can be relied on.

   Without question, such a reputation is adaptive and leads to positive outcomes (even if indirectly) for the individual who chooses to pay it forward.

3. **Loving selflessly.** As we elaborate in material that follows, an enormous body of work from the past two decades on the evolutionary psychology of love (e.g., Fisher, 1993) has demonstrated how strong our love for another can be. This kind of love, which can be selfless, is also an important part of our evolutionary heritage. Human offspring are *altricial* (helpless), and acquiring help from multiple adults (e.g., a monogamous pair of adults) is hugely beneficial to successful development. And when the adults in that pair are fully aligned in their vision of family, which benefits from them being truly in love with one another, parenting will thrive. As described in detail in the next section, love, an inherently selfless act, is a foundational part of the human evolutionary story.

Did Dawkins's juggernaut of a term, *selfish gene*, imply that *all* features of *all* organisms are selfish in the colloquial sense of the term? Absolutely not. He simply meant that the qualities of organisms that lead to gene replication are likely (mathematically) to out-exist qualities that do *not* facilitate such replication. In complex, socially oriented, and long-lived critters like us, it's very often the case that selfless, *other-oriented* behaviors—such as paying it back, paying it forward, or loving another in a selfless manner—are exactly the highly evolved things that make us human, and these are the qualities we share with humans in all corners of the globe.

To some extent, selfish genes in humans have created altruistic apes who focus largely on what they can do to help others and build strong and positive communities. And, of course, this is a foundational aspect of positive psychology.

Why We Love

From an evolutionary perspective, something that is shared widely across a large proportion of a species begs to be explained in evolutionary terms. The nature of a finch's beak must speak to the evolutionary history of finches. The plumage of a peacock's tail must tell us something about how peacocks choose mates and how they ultimately reproduce. The length of a giraffe's neck must tell us something about the kinds of vegetation found in the environments of the ancestors of giraffes. Humans are no different. Features that are universal to our species provide clues to who we are and where we came from. The emotion of love is no exception.

Love is a human emotion that has been documented in human groups across the globe (see Fisher, 1993; Hughes, Harrison, & Gallup, 2007). Further, romantic partners who report themselves as deeply in love consistently show similar neuropsychological activation in cognitive neuroscience studies of the love experience (see Acevedo, Aron, Fisher, & Brown, 2012). The nature of love in humans is a window into our ancestral past.

The evolutionary psychology of love provides an interesting example of how evolutionary principles can be used to shed light on basic aspects of who we are (see Geher, 2014; Geher & Kaufman, 2013).

We Are the Slowly Developing Ape

Species vary in terms of how advanced their offspring are at birth, which ultimately means that they vary in terms of how able their offspring are to care for themselves at birth. Some species are *precocial*, meaning that their off-spring advance relatively quickly. For instance, fawns will get up and start walking on the day they're born.

On the other hand, some species are *altricial*, meaning that their offspring are not particularly advanced at birth, and that they need a lot of time and care in order to develop appropriately. Think about humans: We are not like deer! Our offspring are not walking on their first day. In fact, we're lucky if our offspring are walking in the first year. And even then, anyone who has ever watched a 1-year-old knows full well that they need you to be there every step of the way. Humans are a classic altricial species: We are a slowly developing ape.

Parental Investment and Human Mating Systems

In a classic theoretical piece in the evolutionary sciences, Robert Trivers (1972) developed *parental investment theory*, the idea that the amount of parental investment required in a species should map onto the social and mating-related behaviors of that species. If a species is relatively precocial, we would not expect long-term mating to evolve. Trivers was correct in this prediction across a wide range of species. When you consider a species with quickly advancing offspring, you do *not* find long-term mating, monogamy, or anything of the kind. Bucks and does, for example, spend very little time together outside of mating season

Further, in species with relatively altricial offspring (e.g., birds like the emperor penguin or the North American robin), long-term mating systems may well occur. This is because, in an altricial species, having multiple parents around to help provide resources and raise the offspring can be critical (see Trivers, 1972). This pattern is often called *biparental care*, and it's a hallmark of species with altricial young. You can probably see where this is going: Yes, humans have altricial young, so humans have long-term mating systems and things like monogamy.

The Evolutionary Function of Love

Love within a pair bond seems to be an evolved product of high levels of parental investment in humans (see Fisher, 1993). Love is marked by psychological processes such as passion and intimacy with a specific partner. It is also marked by physiological processes, such as increased levels of oxytocin and autonomic nervous system arousal, which are specific to being in proximity to one's partner (see Acevedo et al., 2012). Love motivates you to be near your partner, to be with your partner, to help your partner, to be kind to your partner, and all of these things make so much evolutionary sense when you consider them in terms of biparental care. Offspring with two doting (and cooperative) adults around to help them simply have an advantage over offspring with only one doting adult around. Love evolved to provide the emotional framework for maintaining pair bonds largely because we are an altricial species with relatively helpless young. Given these ecological factors, childrearing from multiple adults is essential, and a common system for such parenting is found in some variant of monogamy.

Implications for Positive Psychology

When it comes to the emotional and social worlds of humans, understanding the evolutionary history of our emotional system and appreciating the fact that reciprocal altruism runs deep in our species have enormous implications for understanding the positive aspects of human psychology. Among these implications are the following:

- Cultivating happiness is only a proximate goal of life.
- Negative emotions, such as anxiety, anger, and depression, have important functions in our emotional lives.
- All-out selfish approaches to life have large long-term costs.
- Developing a reputation as an altruist has substantial long-term benefits.
- Apologizing and forgiving others is a central feature of human social behavior, each with the evolutionary function of helping people stay connected to their broader group.
- Love is a classic evolutionary adaptation in humans.
- Human pair bonding stems from the fact that our offspring are highly altricial; pair bonding and love help connect parents to help with childrearing.
- Across a range of domains, human emotional and social psychology evolved to help keep individuals strongly and deeply connected to others.

Bottom Line

To best understand the positive emotional and social aspects of human psychology, we need to think about things from an evolutionary perspective. Positive psychologists often focus on figuring out how to increase human happiness. From an evolutionary perspective, cultivating happiness is only a proximate goal. Happiness, like all of the basic human emotions, evolved for specific evolutionary reasons. Understanding the evolutionary origins of our emotion system can help shed important light on how to best cultivate the positive features of our psychological worlds.

From a social perspective, humans evolved with a deep history of reciprocal altruism: Helping others with an implicit understanding of help in return is a foundational aspect of who we are. This fact shapes our emotional and social worlds in important ways. Complex emotions and social behaviors, such as guilt and forgiveness, evolved as mechanisms for keeping individuals connected to others in their social worlds in light of transgressions that take place within social communities. Similarly, love is a social emotion that keeps individuals connected to one another to help effectively raise a family. This broad-based understanding of human social and emotional processes from an evolutionary perspective sheds light on the foundation of positive psychology.

Acknowledgments

Some content from this chapter was adapted from Glenn Geher's (2017) *Psychology Today* blog post "Evolved for Happiness"; his (2014) post "Evolved to Pay It Back and Pay It Forward"; his 2017 post "Helping Your Neighbor"; and his 2016 post "Why We Love." Glenn owns the copyright to the material.

References

Acevedo, B. P., Aron, A., Fisher, H. E., & Brown, L. L. (2012). Neural correlates of marital satisfaction and well-being: Reward, empathy, and affect. *Clinical Neuropsychiatry, 9*, 20–31.

Algoe, S. B., Haidt, J., & Gable, S. L. (2008). Beyond reciprocity: Gratitude and relationships in everyday life. *Emotion, 8*, 425–429.

Chopik, W. J., Newton, N. J., Ryan, L. H., Kashdan, T. B., & Aaron, J. (2017). Gratitude across the life span: Age differences and links to subjective wellbeing, *The Journal of Positive Psychology*. doi:10.1080/17439760.2017.1414296

Darwin, C. (1872). *The expression of the emotions in man and animals*. London: Murray.

Dawkins, R. (1976). *The selfish gene*. Oxford: Oxford University Press.

Fisher, H. (1993). *Anatomy of love—A natural history of mating and why we stray*. New York: Ballantine Books.

Geher, G. (2014). *Evolutionary psychology 101*. New York: Springer.

Geher, G. (2014). Evolved to pay it back and pay it forward. *Psychology Today* blog.

Geher, G. (2016). Why we love. *Psychology Today* blog.

Geher, G. (2017). Evolved for happiness. *Psychology Today* blog.

Geher, G. (2017). Helping your neighbor. *Psychology Today* blog.

Geher, G., & Kaufman, S. B. (2013). *Mating intelligence unleashed*. New York: Oxford University Press.

Guitar, A, E., Glass, D. J., Geher, G., & Suvak, M. K. (2018). Situation-specific emotional states: Testing Nesse and Ellsworth's (2009) model of emotions for situations that arise in goal pursuit using virtual-world software. *Current Psychology*. https://doi.org/10.1007/s12144-018-9830-x

Hamilton, W. D. (1964). The genetical evolution of social behaviour. *International Journal of Theoretical Biology*, 9, 1–16.

Hughes, S. M., Harrison, M. A., & Gallup, G. G., Jr. (2007). Sex differences in romantic kissing among college students: An evolutionary perspective. *Evolutionary Psychology*, 5, 612–631.

Kashdan, T., & Biswas-Diener, R. (2014). *The upside to your dark side*. New York: Avery.

McMahan, E. A., Choi, I., Kwon, Y., Choi, J., Fuller, J., & Josh, P. (2016). Some implications of believing that happiness involves the absence of pain: Negative hedonic beliefs exacerbate the effects of stress on well-being. *Journal of Happiness Studies*, 17, 2569–2593. doi:10.1007/s10902-015-9707-8

Nesse, R. M., & Ellsworth, P. C. (2009). Evolution, emotions, and emotional disorders. *American Psychologist*, 64, 129–139.

Nesse, R. M., & Williams, G. C. (1995). *Why we get sick: The new science of Darwinian medicine*. New York: Times Books.

Trivers, R. L. (1971). The evolution of reciprocal altruism. *Quarterly Review of Biology*, 46, 35–57.

Trivers, R. (1972). Parental investment and sexual selection. In B. Campbell (Ed.), *Sexual selection and the descent of man*: 1871–1971 (pp. 136–179). Chicago: Aldine.

Trivers, R. (1985). *Social evolution*. Menlo Park, CA: Benjamin/Cummings.

Watkins, P. (2014). *Positive psychology 101*. New York: Springer.

7

Taking the High Road in Life

In integrating ideas from the field of evolutionary psychology to illuminate questions connected with positive psychology, this book deals with a constant tension surrounding the human experience. Humans are the products of organic evolution, and we have a suite of adaptations that facilitate our own survival and reproductive success. Yet humans have a unique social ecology, having evolved under conditions in which we interacted with the same individuals, often including nonkin, over and over again across relatively long life spans. Humans, then, evolved a unique combination of features that, on one hand, advance one's own immediate interests and that, on the other hand, subordinate one's own interests in deference to others at the same time.

This bifurcate portrait of the human experience opens the door to multiple pathways in life. One path is the path of selfishness, a path that corresponds to many of our evolved proclivities at a basic level. Another path is a path dominated by self-sacrifice, a path that corresponds to adaptations connected with the unique kind of group living that ancestral hominids experienced in early human groups.

This chapter is all about the tension between these two broad approaches to living.

Multiple Strategies to Success in Life

Evolutionary psychologists are keen on the idea of *strategic pluralism* (see Gangestad & Simpson, 2000), the idea that there are multiple routes to evolutionary success. For instance, in many species, such as sunfish, two basic male mating strategies exist: For one, males who are large and dominant attract females by displaying these qualities and by hovering over a high-quality nest near the water's edge. An alternative strategy is a *sneaky* strategy in which a smaller male takes on the behavioral qualities of an unassuming and nonthreatening female waiting nearby the nest of a large male. Such sneaker fish will strike by releasing gametes when a real female comes to mate with the dominant male—getting its sperm to activate the eggs of the female.

Research on reproductive success in sunfish shows that each of these two basic strategies is effective. Such strategic pluralism is found in many species along many behavioral domains.

In humans, there are all kinds of variability in behavioral strategies. A basic way to think about such strategic variability in humans pertains to the concept of life history strategy (see Figueredo et al., 2005). The idea of life history strategy is that there are two basic approaches to living that an organism might follow. Importantly, the approach that is likely to be utilized depends largely on environmental conditions. Under stable and resource-rich conditions, an organism can afford to take a *slow* approach to life. In terms of work on the topic of life history strategy, such a slow approach will often play out in terms of delayed reproduction and having more offspring, each of which gets a high amount of attention and energy. If life is good, you can take it slow.

On the other hand is a *fast* life history strategy. Such a strategy tends to correspond to signals of relatively unstable and resource-depleted conditions. If the environment is unpredictable and lacking in resources, then a slow approach to life might not be optimal. Under such conditions, organisms are likely to pursue a relatively fast approach to life, reproducing relatively early, having more offspring, and spending less attention and energy on any particular mate or offspring. If life is unstable and unpredictable, you need to move fast.

Dark Personality Traits

A related way to think about variability in behavioral strategies from an evolutionary perspective pertains to what we call *dark* personality traits. The Dark Triad (see Jonason, Kaufman, Webster, & Geher, 2013) includes three seemingly unrelated traits that tend to cluster together. These traits are *Machiavellianism*, defined as using others for one's own gain; *narcissism*, defined as being overly focused on oneself; and *psychopathy*, which is characterized by a strong tendency to have little regard for the feelings of others. Interestingly, scoring high on any one of these three *dark* traits corresponds to scoring high on either of the other traits. That is, these traits seem to cluster together. From an evolutionary perspective, this fact is a clue suggesting that while scoring high in any one of these dark traits might be maladaptive, scoring high on all three might, actually, correspond to a successful route to social power. Interestingly, scoring high on psychopathy and Machiavellianism corresponds to utilizing a fast life history strategy (see Jonason et al., 2017). So it seems that utilizing a fast life history strategy and scoring high on the Dark Triad go hand in hand.

Evolutionary forces often bring traits together if they, in combination, catalyze one another and help bring about success related to survival and reproduction. Someone who is high in only one of these three traits may not do that well, but someone who is high in all three of these traits seems to achieve social successes via dark means. In other words, being high in all three of the subfacets of the Dark Triad seems to be the key to its success.

A theme in this book focuses largely on the trade-offs between a focus on oneself and a focus on others in one's social world. And it turns out that both approaches can lead to success in life.

The Evolutionary Psychology of Jerks

The idea of a plurality of behavioral strategies, with some focusing on benefiting oneself in a short-term manner and others focusing on taking a slow approach to life and being relatively other oriented, opens the door for understanding the evolutionary psychology of these two disparate approaches to social life. The evolutionary perspective can, in fact, help us understand factors associated with taking a high-road approach to living versus, well, simply, being a selfish jerk!

And let's face it: There are some jerks out there. As an example of the kind of behaviors that we might find in someone who scores high in the Dark Triad, let's think about someone who regularly bad-mouths others. We all know a bad-mouther: the person who says something nasty about nearly everyone at the office, the member of your extended family who insults everyone regardless of relatedness, or the guy in your local community who capitalizes on every opportunity to share how stupid, inept, and hypocritical someone else is. Bad-mouthers are out there and they have teeth.

As members of a species that so strongly values trust, agreeableness, and reciprocal altruism (see R. L. Trivers, 1971), it makes you wonder: How do these people get away with it? What is it about the social strategy of bringing others down that *works*?

The Basic Elements of Bad-Mouthing

The socially strategic foundation of bad-mouthing is to bring others down and to create an uneasy environment. If Joe always casts insults on half the people in his workplace, then you'd better be careful around him lest you become his next target. This behavioral vigilance that Joe creates in others empowers him, potentially allowing him to have an outsize influence on how things go. Joe's power may largely stem from fear and intimidation, an approach to social interactions that is rooted in Machiavellianism.

Why Bad-Mouthing Exists

Bad-mouthing can only endure if it's effective—leading to beneficial social outcomes for the bad-mouther. And, for better or worse, a great deal of research has shown that Machiavellian behavior such as bad-mouthing often does lead to success in various domains, such as the worlds of mating or the workplace (see Geher & Kaufman, 2013).

Following are three reasons that a bad-mouthing social strategy exists, in spite of its obviously unpleasant nature:

1. **Bad-mouthing is a route to social power.** By gaining a reputation as someone who will throw his or her own mother under the bus, a bad-mouther can gain social power via creating a fearful environment. It's socially risky to mess with bad-mouthers, and they capitalize on this fact.
2. **Bad-mouthers exude confidence, a basic catalyst to social success.** Confidence leads to success across a variety of life domains (see Geher & Kaufman, 2013), often regardless of whether it is warranted. And a strategy of putting others down often goes hand in hand with conspicuous displays of confidence.
3. **Bad-mouthers may find themselves in leadership positions.** Putting others down as a strategy toward benefiting oneself may well turn up leadership opportunities, which increase the power of the bad-mouther.

Kinder Routes to Social Success

There are other paths to success: Kind qualities, for example, are consistently found to be as important in social partners, evident in human populations across the globe (see Buss, 2003). Empathy, for instance, is consistently associated with more supportive friendships (Ciarrochi, Sahdra, Kashdan, Kiuru, & Conigrave, 2017). In fact, various forms of prosocial behavior and attributes related to appreciating and supporting others are foundational to who we are (see Geher, 2014).

Does it pay to be someone who gains social power by creating a social world of fear and intimidation? It can. But are there other routes to social success, such as building others up rather than tearing them down? Absolutely.

The Elements of Taking the High Road

This chapter is about taking the high road in life. We see taking the high road as consistent with the goals of positive psychology—taking an approach to life that contributes positively to community works toward the great good and advances one's own physical and mental health along the way.

To this point in the chapter, we have focused on understanding why dark approaches to social life have the capacity to lead to success. In this section, we discuss the flip side to this equation. How and why can prosocial, other-oriented approaches to life lead to success? Next discussed are three broad ways that a bright (rather than dark) approach to others can lead to success in life.

Kindness Is Attractive

One clue to the idea that an other-oriented approach to living can lead to success in one's own life is found in research on mating preferences conducted by David Buss and his colleagues (see Buss, 2003). In large-scale, cross-cultural research examining the qualities that are desired in mates by both males and females, Buss found that there are several differences in the preferences of each gender. For instance, across cultures, males seem to focus more on markers of health and youth, while women seem to focus more on markers of status and resources.

This said, an often-overlooked aspect of Buss's research is this: Both males and females, across the globe, care a great deal about kindness in potential mates—and kindness actually seems to matter more for males and females than do other factors, such as physical attractiveness or wealth. Think about that.

When it comes to evolution, mating preferences exert strong pressures on qualities that come to characterize members of a species. If both males and females are looking for kind mates, then you can bet that kindness has been selected in our species.

So one basic reason that taking an other-oriented approach to life, being kind to others, is rooted in the evolutionary foundational processes of sexual selection—kindness helps one obtain mates.

Altruists Are Trusted by Others

Who would you trust more, the person at work who bad-mouths everyone at the drop of a hat or the most altruistic person in the office? The answer is obvious. In a world that is characterized by large-scale reciprocal altruism, as is found in the human social ecology (see R. Trivers, 1985), reputations matter. So being kind and helpful to others in your social world has both short- and long-term potential consequences. In the immediate sense, helping another may well lead to having that person help you in return. In a longer term sense, helping others may well lead to the cultivation of a reputation as someone who is other oriented and as someone who can be trusted within the community.

An Other-Oriented Approach May Lead
to Leadership Opportunities

Sure, as suggested in the previous section, a dark, nasty, and selfish approach to one's social world might lead to power and to leadership opportunities. But remember that, at the end of the day, humans have proclivities toward egalitarianism (Bingham & Souza, 2009). And in egalitarian, democratic systems, people have opportunities to choose leaders. Of course, there are famous cases of individuals who follow dark paths when emerging in positions of power and leadership (think Richard Nixon). But there are also famous cases of individuals who truly look out for the good of the broader community who end up emerging as leaders (think Barack Obama). If you eventually want to hold some kind of leadership position and have the capacity to make positive changes in your broader community, pursuing an other-oriented approach to life can get you there.

Forgiveness, Estrangements, and Social Life

Just as altruism cuts across many of the chapters of this book, forgiveness too rears its head frequently. In human social life, bad things happen. People sometimes treat each other with disrespect. Sometimes people lie, cheat, or steal. In various ways, people trespass on one another—sometimes intentionally and sometimes incidentally.

In a social world of 150, which, as described previously (see Dunbar, 1992), captures what human social worlds were like for the lion's share of our evolutionary history, you'd better be careful about trespassing on others. What goes around comes around, and that's particularly true in small, tight-knit communities.

An extreme outcome in social living pertains to estrangement, which is characterized by concluding that someone is *dead to you*—indicating that there will be no communication between the two of you ever again.

Occasionally, we all hear people say, "I cut that person off. That person's dead to me. I will never talk to him again." Sometimes a person feels so slighted by another individual that that person is no longer to be treated as a person. Of course, cutting someone off can wreak havoc on many social lives, as shown by the following scenarios:

- Oh, *he* is invited to the party? How could *Steve* have invited *him*! Steve knows what he did to me! Not only am I not going to Steve's party, I'm not inviting Steve to my next party. I may not even stay friends with Steve!
- Don't make me sit next to Sally at the meeting, please! Ever since what happened, she no longer talks to me, and it's always uncomfortable.
- I forbid you to see your cousin Frank! You know what he did last summer—that was such an insult to me and to all of us! *So* disrespectful. How could you possibly still love me and remain close with him? You have to choose!

The Evolutionary Reasoning of the Cutoff

Social ostracism of any kind is difficult and uncomfortable to manage. It makes for difficult social relations. But it was worse during ancestral times. For the lion's share of human evolution, social worlds rarely exceeded 150 people, and folks had to deal with the same 150 people each and every day (see Dunbar, 1992). Today, you could pick up, move to Chicago, and start anew. That kind of thing just wasn't an option under ancestral conditions.

Historically, getting kicked out of one's band would have caused the most dire of consequences. Being cut off from a small number of folks in a group of 150 could easily lead to being cut out by a larger subset of individuals over time. Having important social connections removed could have meant death or a lack of reproductive opportunities—both of which are evolutionary dead ends.

Human social psychology is thus highly sensitive to markers of social alienation. Signs that one is *cut out* from others are signs that create disproportionate levels of social anxiety.

Given this evolved psychology (see Geher, 2014) that we all share, cutting someone out of one's social world can be an effective strategy at making someone feel really bad about himself. Cutting others off is a social strategy that plays off our evolved psychology.

Forgiving Instead of Cutting Someone Off

Forgiveness is an important behavior related to dealing with such social situations. When someone feels slighted and expresses outrage as a result, the slighter, who may feel shame and remorse, will often take steps to seek forgiveness. In the world of positive psychology, forgiveness is seen as a character strength. Forgiveness has been shown to minimize stress and increase mental health (Toussaint, Shields, Dorn, & Slavich, 2016), so people struggling with these are often encouraged in a clinical setting to foster more forgiving coping styles if that's a character strength that is otherwise lacking.

There are important benefits to forgiving. First, forgiving others has the potential to raise one's reputation as being other-oriented. When done carefully and in a way that doesn't make one look like a punching bag in the broader group, it's a signal that one is kind and highly trustworthy—and that one has the interests of the broader social group at heart. These are all qualities that we value in others, especially those with leadership positions. Research also shows that it feels good to forgive (see Gorsuch & Hao, 1993), suggesting that there must have been real benefits to our ancestors who were forgivers.

The social cutoff is one of the most difficult things to deal with. From an evolutionary perspective, we can understand why people implement it. But an evolutionary perspective also sheds light on a more productive approach to dealing with being slighted in social situations. When done well, forgiveness ends up not only keeping a social circle intact, but also has the capacity to raise the status level and respect that people feel for the forgiver. There are good evolution-based reasons for the belief that to forgive is divine.

The High Costs of Estrangements

A recent study conducted by the New Paltz Evolutionary Psychology Lab speaks to the high social and emotional costs of estrangements (Geher et al., 2017). This study took an evolutionary approach to social estrangements and predicted that a higher number of estrangements in one's life would be associated with a broad suite of adverse social and psychological outcomes. We surveyed over 300 adults and asked them to demarcate the number of estrangements that they had in their lives. The number of estrangements in our sample ranged between 0 and 27.

We also had participants complete measures of social and psychological well-being, including indices of emotional stability, social support, well-being, and depressive tendencies. We also measured the Dark Triad.

In our analyses, we divided participants into two broad categories: very high in the number of estrangements (10+) or not (9 or less). Consistent with our hypotheses, we found very strong and consistent effects associated with having a very high number of estrangements. Individuals with 10 or more estrangements scored as high on the Dark Triad, emotional instability, and depressive tendencies. They also scored as being insecure in terms of their attachments with others. It seems that being cut off from a large number of others does, indeed, have severe and broad-ranging consequences for one's life. This is all the more reason to try to cultivate an approach to life that is founded on mutual respect and inclusivity.

Trespassing, Apologizing, Forgiving, and Moving Forward

In a sister study to the estrangement project, the New Paltz Evolutionary Psychology Lab has also recently collected data on factors associated with forgiving someone in your social world (Geher et al., 2018). This study explored three factors as they related to forgiving a trespass or an insult in one's social world. These factors were (a) intensity of the trespass (very intense or benign); (b) target of the trespass (at oneself or at one's property); and (c) whether the trespasser apologized. Using an experimental design with eight different conditions, we created two sets of stimuli regarding hypothetical social situations. One set included your friend *Victor*, who said something behind your back at a party, and the other included *Lauren*, who unwittingly insulted you at a restaurant.

One example vignette is as follows:

Your friend Victor and you were talking at a party that you were hosting. The doorbell rang and you excused yourself briefly. Just as you came back into the room, you overheard Victor say **that he thinks you're such a joke and he just came for the food.**

You immediately catch Victor's eye at that moment. Victor then says to you, **"No offense, but it's true."**

Here, we have the situation in which the trespass was high in intensity (*you're such a joke*) and at the target him- or herself (as opposed to at the target's property, as in some of the other conditions), and there was no apology offered.

After participants read each of the vignettes to which they were randomly assigned, they responded to several items on a 1–7 scale indicating how they would feel in this scenario. Items included their level of anger, the degree to which they would forgive the person, and the degree to which they would remain friends with the person after the incident.

Results were intriguing. The two factors that had the largest effects were the intensity of the insult along with if it was directed at the person or at the person's property. For pretty much all of the dependent variables, these factors mattered a great deal. Interestingly, the apology variable only played a role in the question of whether the participants would remain friends with the insulter.

Further, for all participants, we measured the Dark Triad, and the results were interesting. Regardless of the experimental condition that one was in, participants who scored relatively high on each facet of the Dark Triad were (a) less likely to forgive, (b) less likely report wanting to stay friends, and (c) likely to seek revenge. These results put a pretty specific face to the inner psychology of someone with a dark personality.

Overall, the lessons of this study dovetail strongly with the lessons of the estrangement study. It's important for humans to stay in the good graces of others. From an evolutionary perspective, in fact, it's essential. If you want to succeed in life, follow the Golden Rule and treat others as you'd like to be treated. And when problems emerge, genuinely apologizing may help maintain friendships in spite of difficulties.

Implications for Positive Psychology

Once we start thinking about what it means to take the high road from an evolutionary perspective, we can begin to understand the basic features of taking the high road in social interactions. All of these ideas can help inform the field of positive psychology. Specific implications for positive psychology are as follows:

- Try not to use dark strategies to succeed—avoid Machiavellianism, narcissism, and a callous, uncaring (psychopathic) approach to others.
- Don't talk behind the backs of others—such a strategy has at least as many social costs as benefits.
- Kindness will get you far—in mating as well is in all other kinds of social domains.
- Humans evolved to be altruistic in many ways—never forget that we are the altruistic ape.
- Remember that humans evolved for small-scale living—so theories designed to help us live better lives need to take this critical feature of our ancestral past into account.
- Social estrangements wreak havoc on individuals and communities.
- Insulting others or trespassing on others wreaks havoc on relationships.
- Moral emotions and behaviors, such as forgiveness and apologizing, go a long way toward restoring personal and interpersonal problems.
- Watch out for those in your social world who show the hallmarks of being high in the Dark Triad.

Bottom Line

Life is full of trade-offs that deal with costs and benefits to oneself versus to one's group. In an immediate sense, it's easy to take the selfish path in any situation, but such an approach is not without costs. From an evolutionary perspective, we can understand how taking the high road in life—by not talking badly about others, being kind and other oriented, being forgiving, and apologizing when appropriate—has the capacity to help provide positive outcomes for individuals as well as positive outcomes for one's community. While it may not always be easy, taking the high road in life pays dividends in the long run.

Acknowledgments

Some content from this chapter was adapted from Glenn Geher's (2015) *Psychology Today* blog post "The Psychology of the Badmouther" and his 2015 post "What It Really Means When Someone Says, 'Dead to You.'" Glenn owns the copyright to the material.

References

Bingham, P. M., & Souza, J. (2009). *Death from a Distance and the Birth of a Humane Universe*. Charleston, South Carolina: Booksurge.

Buss, D. M. (2003). *The evolution of desire: Strategies of human mating*. New York: Basic Books.

Ciarrochi, J., Sahdra, B. K., Kashdan, T. B., Kiuru, N., & Conigrave, J. (2017). When empathy matters: The role of gender and empathy in close friendships. *Journal of Personality*, *85*, 494–504.

Dawkins, R. (1976). *The selfish gene*. Oxford: Oxford University Press.

Dunbar, R. I. M. (1992). Neocortex size as a constraint on group size in primates. *Journal of Human Evolution*, *22*, 469–493.

Figueredo, A. J., Vásquez, G., Brumbach, B. H., Sefcek, J. A., Kirsner, B. R., & Jacobs, W. J. (2005). The K-factor: Individual differences in life history strategy. *Personality and Individual Differences*, *39*, 1349–1360.

Gangestad, S. W., & Simpson, J. A. (2000). The evolution of human mating: Trade-offs and strategic pluralism. *Behavioral and Brain Sciences*, *23*, 573–644.

Geher, G. (2014). *Evolutionary psychology 101*. New York: Springer.

Geher, G. (2017). What it really means when someone says "dead to you." *Psychology Today* blog.

Geher, G. (2017). The evolutionary psychology of the badmouther. *Psychology Today* blog.

Geher, G., Baroni, A., Nitza, E., Sullivan, G., Dobosh, K., & Stewart-Hill, S. (2018, April). *I just came for the food. And no, I'm not sorry!* Presentation given at the annual meeting of the NorthEastern Evolutionary Psychology Society, New Paltz, NY.

Geher, G., Betancourt, K., Chason, M., Eisenberg, J., Holler, R., Mabie, B., . . . Gleason, M. (2017, June). *You're dead to me: The evolutionary psychology of estrangements*. Presentation given at the annual meeting of the NorthEastern Evolutionary Psychology Society, Binghamton, NY.

Geher, G., & Kaufman, S. B. (2013). *Mating intelligence unleashed*. New York: Oxford University Press.

Gorsuch, R. L., & Hao, J. Y. (1993). Forgiveness: An exploratory factor analysis and its relationship to religious variables. *Review of Religious Research, 34*, 351–363.

Hamilton, W. D. (1964). The genetical evolution of social behaviour. *International Journal of Theoretical Biology, 7*, 1–16.

Jonason, P. K., Foster, J. D, Egorova, M. S., Parshikova, O., Csathó, Á., Oshio, A., & Gouveia, V. V. (2017). The Dark Triad traits from a life history perspective in six countries. *Frontiers in Psychology: Evolutionary Psychology, 8*, 1476.

Jonason, P. K., Kaufman, S. B., Webster, G. D, & Geher, G. (2013). What lies beneath the Dark Triad Dirty Dozen: Varied relations with the Big Five. *Individual Differences Research, 11*, 81–90.

Jonason, P. K., Lyons, M., & Blanchard, A. (2015). Birds of a "bad" feather flock together: The Dark Triad traits and mate choice. *Personality and Individual Differences, 78*, 34–38.

Miller, G. F. (2000). *The mating mind: How sexual choice shaped the evolution of human nature*. London: Heineman.

Toussaint, L., Shields, G. S., Dorn, G., & Slavich, G. M. (2016). Effects of lifetime stress exposure on mental and physical health in young adulthood: How stress degrades and forgiveness protects health. *Journal of Health Psychology, 21*, 1004–1014.

Trivers, R. (1985). *Social evolution*. Menlo Park, CA: Benjamin/Cummings.

Trivers, R. L. (1971). The evolution of reciprocal altruism. *Quarterly Review of Biology, 46*, 35–57.

Wilson, D. S. (2007). *Evolution for everyone: How Darwin's theory can change the way we think about our lives*. New York: Delacorte Press.

8

Evolutionary Psychology of Resilience

Let's start with a story, a tale about an ugly duckling—you know the one. Hatched into a cruel world, the ugly duckling wanders around his barnyard and endures both verbal and physical abuse. Other animals tease him because of his homely appearance. He flees to go live among wild geese, but hunters slaughter the flock. The duckling then finds a home with an elderly woman, but her cat and hen also taunt him mercilessly. He sets off alone once again. On the ugly duckling's lonesome journey, he discovers a beautiful flock of swans. To his dismay, he is too young to fly among them. Winter arrives and his lonesome journey continues. A helpful farmer finds the ugly duckling and carries him home to keep him warm, but the farmer's children scare the duckling. He flees and spends the winter alone outdoors. The poor guy just can't catch a break. When spring arrives, the flock of beautiful swans returns. The ugly duckling tries, yet again, to join the swans—determined little guy. He refuses to give up and live his entire duck life alone while subjecting himself to such cruelty. You can guess what happens next. Much to his surprise, the ugly duckling is accepted by the swans. In fact, they more than accept him—they welcome him with great zeal! The ugly duckling then sees his reflection in the water to find that he has grown into one of the most beautiful swans of all. He lives happily ever after. Probably ...

Resilience—Everyone's Doing It

We all know this story. It's a relatable tale meant to remind us to carry on. To keep at it. Don't give up the fight. The world is not a very forgiving place, and life likes to remind us of that every now and again. But here's the cool part— we *do* carry on. It's called *resilience*. Resilience can be described as a rebound. It is an ability to bounce back in the face of challenge or hardship, and it isn't limited to ugly little ducks. Humans do this, too.

Heck, one could probably argue that if we weren't resilient, we might never find love. If we gave up on relationships after our first dramatic and oh-so-intense high school breakup, we'd never find *the one*. Resilience is what leads us to a happy relationship after multiple failed ones. It's what leads us to our dream job after multiple underpaid and failed ones. It's the couple of steps we take forward after life sets us one back. Without resilience, life would be pretty darn miserable or, at the very least, mundane.

One Step Back and Two Steps Forward

Resilience is not only about pushing forward when you have a head start and are on the right track (see Masten, 1989). It really comes into play in the face of adversity—when you're starting closer to rock bottom. A powerful example of resilience can be seen when we look at the attack on the twin towers on September 11, 2001. It remains the single largest attack on American soil to date, with roughly 3,000 casualties. In the face of such a devastating catastrophe, the American people came together to cope and rebuild. People donated time and money and labor in an effort not only to restore what had been, but also to come out stronger than before.

On a smaller scale, we know that resilience is also a strong force on an individual level. Several longitudinal studies have shown that despite enduring various forms of hardship and abuse, people cope and persevere (Werner, 2012). Anyone reading this right now likely can reflect on personal moments of resilience and strength. I encourage you to think back to a time of struggle or pain, maybe an accident of some kind, the death of a loved one, the loss of a job, the loss of a pet, a breakup, etc. Did life end for you there? Or did you keep on keepin' on? What encouraged you to find the silver lining and come out stronger than before? We're stronger and more resilient than I think we give ourselves credit for. I've heard people say, "Oh gosh, I don't know how I could possibly deal with X like so-and-so did," a million times. You never know until you have to deal with X, but we find a way. One foot goes in front of the other—left, right, repeat. Breathe in, breathe out, repeat. Over time, we heal and come out stronger than before. Ironic (and perhaps unfortunate) as it may seem, failure and hardship are actually the biggest predictors of future success.

Try and Try Again

If you're anything like we are, you can easily sit down at the computer in front of a blank page titled *My Failures in Life* and go to town. In our backgrounds, you would find summaries of rejections from colleges, rejections from graduate schools, rejections from publications, rejections from job applications, embarrassing breakups, and lots of participation ribbons and medals. And that's just the start.

I (N. W.) played one season of basketball in the third grade. Before the last game of the entire season, my coach brought the team in for a huddle. The plan for the game was, "Any time you have the ball, pass it to Nicole! She's the only one who hasn't scored a single point all season." Yikes. And just to answer your next question, no, I did not score in that game either. I'll admit right now that I did not display a ton of resilience when it came to basketball in particular. It's just never going to be one of my strengths. But I *did* display resilience with sports in general. I didn't totally give up after the third grade; I went on to try gymnastics, Taekwondo, soccer, and horseback riding, and I turned out to be a decent musician. All of these were better suited for me than basketball, and I've learned so much from each sport and experience. Imagine if I'd have just thrown in the towel (pardon the pun) after third-grade basketball—I would have missed out on so much! I also probably wouldn't have broken my nose so many times, but that's another story for another book and my plastic surgeon.

Don't Give Up. Don't Ever Give Up

As academics, our primary goal is to help bright young minds develop. As such, we are interested in the success of our students, and of course we don't ever want our students to give up—ever. In 1993, the famed North Carolina State men's basketball coach Jimmy Valvano (also known as Jimmy V.) looked a national ESPN audience in the eye and, his body riddled with cancerous tumors, said this, "Don't give up. Don't ever give up."

He may not have been a research psychologist with a PhD, but Valvano knew what he was saying. And as humans who are capable of resilience, we all have a little Jimmy V. somewhere inside us.

Why was Valvano right? Why should we stand up in the face of adversity? Is this really an effective behavioral strategy? Was it adaptive for our ancestors as they dealt with all kinds of threats on the preagrarian savanna? Heck, yes, Jimmy V. was right! And there is a lot of work in the field of psychology to support his message. The following are five scientifically documented reasons to endorse a Jimmy V. approach to life:

1. **An overly optimistic take on oneself is adaptive.** Much research in psychology (see Krueger, 1998) showed that people generally take a rosy-glassed approach in perceiving themselves. Interestingly, people who are more likely to show such self-enhancement in their self-perceptions are on a track for success in multiple domains.

2. **Having an illusion of control is adaptive.** In reality, we can only control so much of what happens in our worlds. But people vary in terms of how much they tend to *think* they have control—regardless of whether they actually have it (see Presson & Benassi, 1996). The kicker is this: People who think they have a little more control than is actually warranted are at a dramatically reduced risk for depression. And this is a must-have evolutionary-based value. Our ancestors who continued to try at some task—partly because they thought they could control the situation—had to have been our ancestors who (at least sometimes) succeeded because they kept at it.

3. **Think you can! Think you can!** Self-efficacy (see Bandura, Barbaranelli, Caprara, & Pastorelli, 1996) is a pretty straightforward but powerful psychological phenomenon. It's the simple belief that you can accomplish some task. Suppose you need to unlock a rusty old door, and in front of you, I pick a key from a pile of hundreds of keys and say, "Try this one—it might work." Well, you could try it, but you probably would be so doubtful that you might, unwittingly, not exactly give it your all. Imagine another scenario in which I give you a single key, and I say, "This is definitely the key." Well you might jiggle it and such, but you'd probably try harder than you might in the other condition—even if the keys were identical. This is how it works: If you think you might succeed, you try harder—and such effort often begets success. Whenever possible, shoot to foster self-efficacy in all your endeavors. It's sort of Step 1 for success.

4. **Have an overly optimistic take on others in your life.** Humans don't live in vacuums. We live in specific social circles. We have others who comprise our primary support group—often our spouse, family, and close friends. People often extend the self-relevant biases described to

these close individuals. For instance, people tend to overidealize their romantic partners (see Geher et al., 2005). In fact, overidealizing one's romantic partner is a huge predictor of relationship success and satisfaction. Give others in your life the benefit of the doubt and put on some rosy glasses when looking at them!

5. **Life is short. Do something great.** This point may not be as scientifically based, but it's still spot on. Resilience (see Masten, 1989) is a set of qualities that help us fend off the adverse effects of negative outcomes such as failure and rejection. We are all going to face rejection and failure; that is part of life. Our ability to effectively combat such outcomes and bounce back is resilience, and it is how we can take failures in these short lives that we have and turn them, ultimately, into successful outcomes marked by greatness.

If you are a successful human (and if you've read to this point in the book, then you surely qualify), you need to look failure in the eye and rise above it. The most successful people are also often the ones who have experienced the most failure. Failure and rejection hurt, but they are not showstoppers. They are important features of life that help us grow stronger and that help us succeed into the future. When Valvano had a body full of cancer and months left to live, what did he do? He started the Jimmy V. Fund to help raise a ton of money to help facilitate scientific cancer research. I'd call that a success story, actually.

So if you have run into some kind of failure or rejection lately, we say this to you: Foster your belief in yourself, foster your belief in close others who support you, and unleash your inner Jimmy Valvano: *Don't give up. Don't ever give up.*

Resilience as a Biological Adaptation

From a biological perspective, the idea is that some features of organisms are more likely, on average, to make it into future generations compared with other features. Handgrip strength that is strong is likely to lead to better brachiation and tree climbing, increased chance of survival, and more reproduction than weak handgrip (Hello! Ever seen *American Ninja Warrior*?!). But while some features (e.g., strong and versatile handgrip strength) outcompete other features (e.g., wimpy handgrip strength), all will show some level of failure. Adaptations for strong handgrip strength may well have evolved over thousands of generations during the evolution of the arboreal primates in Africa, but we must note that in this process, failure, leading to primates falling from canopy to the forest floor, was necessarily a sometimes consequence. Many a primate fell to a gory death during the years when advanced handgrip strength was evolving. Failure was part of what happened sometimes. We like to think of this example as a species-level sort of resilience. While one primate's weak handgrip strength led to his demise, another primate was able to carry on with a stronger handgrip and reproduce. Members of the species were resilient (thanks to the cold process of natural selection) and continued to evolve as failure continued to pave the way for success. Michael Jordan had it pretty darn right when he said, "I've failed over and over and over again in my life. And that is why I succeed."

Evolutionary processes work this way. They follow a probabilistic logic: Some qualities are "more likely" to lead to success than are other qualities, but they will still have failure rates that are different from zero. A "good adaptation," for instance, may lead to a 10% death rate, while a "not as good" adaptation may lead to a 30% death rate. From this mathematical and evolutionary perspective, *failure is necessarily a part of the game*. The issue is not whether one feature will fail and another will not—the issue is more subtle, nuanced, and statistically based. The issue is whether one feature will, on average, lead to a higher proportion of successes relative to failures compared with alternative features.

Evolutionary Failure and Real Life

All this conceptual stuff about how evolution works has real implications for how our lives progress. In life, you sometimes succeed, and you sometimes fail. This is just how it goes. When we step back and look at organic evolution, the same exact process is true: Some biological adaptations succeed (and come to typify a species), and other such adaptations fail (and come *not* to typify *any* species).

Picking Dandelions to Learn About Success and Failure

But the evolution of life is relentless—and this point needs to be included in this discussion. Ever pick dandelions out of your yard? Good luck. You may start with 20 in your basket, increase to 60, commit to "pick them all," only to find that you have picked 80 after several hours and that 85 more (that you had not seen before) are now in your side yard—and so forth. In the evolutionary story of a modern yard, dandelions show an extraordinary failure rate (they get picked and mowed a lot), only to be out-done by an even more extraordinary success rate (they find good environments and grow a lot).

What happens when you squash a dandelion plant? (a) Does it cry? (b) Does it say, "Go ahead without me—I just cannot do this anymore!" or (c) Does it follow its biological design and disseminate seeds in wayward, random directions, with the (apparent) goal of growing more plants all over the place? If you guessed Answer c, you are correct.

Dandelions, and so many other natural forms of life, have the greatest possible lessons for all of us humans out here—whether we know it or not. And here it is: Dandelions cannot help but fail at times. They don't seem to have evolved mechanisms designed to reduce failure at all! They don't bite; they are actually pleasant to eat (with few if any toxins); they are helpless! Rather, their strategy toward proliferation seems more like this: (a) grow a lot, (b) grow quickly, (c) grow wherever, (d) turn to seed as soon as possible, and (e) go back to Step a. Ever see a field full of dandelions? I bet you have. And that, my friend, is because this particular evolved strategy— resilience—works. It is, on average, effective at facilitating growth and reproduction.

More Failure Corresponds to More Success

The irony of the dandelion is this: The more failure the plants encounter corresponds strongly to the more success that the plants encounter. In essence, these plants are (functionally) *trying*—and they consistently make efforts at replicating. They often get squashed. A 6-year old may decide to make a daisy chain out of them or give a bouquet to a lucky parent. A lawn mower may actually take some of these soldiers down for some time. But in the end, the evolutionary strategy of the dandelion is just so strong. They *always* come back.

Their general plan is simply this: grit and perseverance. Keep at it—move forward through failure, and you are likely to see another day and grow, or at least your offspring will.

Human Success Maps Onto Dandelion Success

Humans are a lot like dandelions. We try all kinds of things. For instance, if you are a kid trying out for a part in a play, you may get rejected the first year (as a dandelion may get weed whacked in Season 1). But the dandelion, due to its biological design, keeps trying. It tries a new area of the yard, or its pods and the wind may bring it a mile away. It gives it another go, and that may work.

Does this strategy work for you? You didn't get this particular part in this particular play. Should you just go belly-up, then? Or maybe you try another play, another role, another venue, another group, another accent. Give something else a try—this may be the solution!

And don't get discouraged. The dandelion may fail five more times before it finally yields a plant that has just the right conditions to facilitate reproductive success. The four or so failures beforehand were just part of the process.

Think about any human domain in which success is a goal. We could learn quite well from the natural world. Suppose you want to successfully publish a scientific journal article. Well, if you ask any scientist you will be told that failure early on in the process is par for the course. Any good scientific article may well have been rejected a solid three other times by other journals before it was accepted, but a good and dedicated scholar knows this—and keeps at it. Just as, with evolutionarily ancient and nonconscious rules, dandelions seem to *decide* where and when they will take up new territory. And, like scientific manuscripts, they will probably fail. But also like the scientific manuscripts of persistent and successful scholars, they will be resubmitted—ultimately to the point of production and publication.

Failure Is a Prerequisite for Success in Life

Those who do not try are those who do not succeed. The most successful among us are, without exception, those who have failed the most—as a result of being those who have tried the most (see the discussion of dandelions, Michael Jordan, or the ugly duckling).

The greatest scholarly successes ever came on the heels of mountains and mountains of failures. It's the perfect picture of resilience. And that is OK. That is natural. The greatest dandelion fields in the world were preceded also by fields and fields of lawn mowers and poor soil conditions and other forms of dandelion adversity. But, based on their evolutionary history, dandelions are like the honey badger: They just don't care! A honey badger doesn't give a [four-letter word]!

Humans who are trying to accomplish something can learn a lesson here from their sisters, the dandelions. Perfectionism has little place in production and optimization. Grit, effort, and persistence supersede perfectionism in cultivating success in many areas—in dandelion reproduction and in human production.

Whatever You Are Doing, Channel Your Inner Dandelion

Be like a dandelion. Try, expect to fail; try, deal with a failure; try, deal with a different failure; then, one day, succeed. You will have a field full of dandelions, a vita full of publications, a classroom full of students who understand the material, or whatever it is that *you* are striving for. Realizing that the failure-to-success ratio in any endeavor is high should go a long way to helping people stay on track and moving toward their goals.

Based on this reasoning, the most successful among us—in any field—are those who have failed the most. And as a corollary, failing a lot is highly predictive of ultimate success and innovation—in any field. This is part of the deal of who we are, and understanding our evolutionary roots helps us get exactly why failure is ultimately, however ironic, predictive of success.

Implications for Positive Psychology

Drawing from our cave-dwelling ancestors, we know that resilience is something that simply *had* to be adaptive going way back into our species' history. There was no other option. Resilient organisms will outsurvive and outreproduce relative to nonresilient conspecifics. Giving up after the first knockout with a competitor, predator, or unsuccessful romance would not have led to where we are now. Understanding that resilience has been a part of the human condition for as long as humans have existed is crucial. It's easier to understand and study resilience from a positive psychology standpoint when we know that it is rooted in our biological past. Resilience helps us to thrive and overcome obstacles, and there are a lot of potential implications

for this. Studies following trauma, crises, natural disasters, and other cata-
strophic events can help to shed light on how resilient we are, as well as what
to practice during times of struggle.

Bottom Line

Resilience is a core feature of human psychology, rooted deep in the ancestral past of all life forms. Cultivating resilience, which is a core goal of the
modern positive psychology movement, is best understand through an evolutionary lens. Look, life is hard. And unlike dandelions, we are capable of
having intensive emotional experiences. Exuding resilience and fighting the
good fight are certainly not always easy. It's never easy to put on a smile and
try to face another day when you're struggling, but it is worth it. Always keep
in mind that you are the product of millions of generations of resilient life
forms that go deep into evolutionary time. And the hard work, grit, and perseverance will be worth it. Just ask the ugly duckling or Jimmy V.

Acknowledgments

Some content from this chapter was adapted from Glenn Geher's (2015)
Psychology Today blog post "5 Reasons You Should Never Give Up" as well as
his (2014) post "Failure as the Single Best Marker of Human Success." Glenn
owns the copyright to the material.

References

Bandura, A., Barbaranelli, C., Caprara, G. V., & Pastorelli, C. (1996). Multifaceted impact
of self-efficacy beliefs on academic functioning. *Child Development, 67,* 1206–1222.

Geher, G. (2014). Failure as the single biggest marker of human success. *Psychology
Today* blog.

Geher, G. (2015). Five reasons you should never give up. *Psychology Today* blog.

Geher, G., Bloodworth, R., Mason, J., Downey, H. J., Renstrom, K. L., & Romero, J. F.
(2005). Motivational underpinnings of romantic partner perceptions: Psychological
and physiological evidence. *Journal of Personal and Social Relationships, 22,* 255–281.

Krueger, J. (1998). Enhancement bias in descriptions of self and others. *Personality and
Social Psychology Bulletin, 24,* 505–516.

Masten, A. S. (1989). Resilience in development: Implications of the study of successful
adaptation for developmental psychopathology. In D. Cicchetti (Ed.), *The emergence of*

a discipline: Rochester Symposium on Developmental Psychopathology (Vol. 1, pp. 261–294). Hillsdale, NJ: Erlbaum.

Presson, P. K., & Benassi, V. A. (1996). Illusion of control: A meta-analytic review. *Journal of Social Behavior and Personality, 11*, 493–510.

Werner, E. E. (2012). *What can we learn about resilience from large-scale longitudinal studies?* (pp. 87–102). New York: Springer.

SECTION III

APPLIED POSITIVE EVOLUTIONARY PSYCHOLOGY

In many ways, positive evolutionary psychology is, as a field, an applied endeavor. This said, Section III is explicitly applied in nature—focusing on specific areas of human functioning that can be improved by applying an evolutionary perspective. The two areas that are the focus in this section are human health and human communities.

With the rise of work in the field of Darwinian medicine, it has become more and more clear that insights from the field of evolution have the capacity to inform issues of human health. In fact, as you will see in Chapter 9, the Darwinian approach to understanding people has led to great advances and new insights in terms of both physical and mental health. If you want to live a healthy life, you ignore ideas inspired by Darwin at your own peril.

Another clear area of application pertains to the nature of human communities. Going back thousands and thousands of generations, *Homo sapiens* have been communal in nature, creating groups of coordinated individuals that often cut across lines of kinship. Building and working with healthy communities have always been foundational features of our species—and the evolutionary perspective presents us with a battery of insights into healthy human communities.

9

Healthy Living Reconsidered

You would think that medical practitioners would have a strong grounding in evolutionary science. In fact, if you thought that, you'd be generally wrong. In their famous exposé on Darwinian medicine, Nesse and Williams (1995) unleashed an entirely new field of inquiry focusing on applications of Darwinian principles to questions of health. Along the way, they provided strong evidence suggesting that Darwin's ideas had been neglected by the field of medicine almost entirely to that point. Given how powerful Darwin's ideas are in illuminating questions related to life, you'd think that the health professions would be all over evolutionary principles. In fact, the application of evolutionary concepts to issues of human health is an extremely recent endeavor.

The advent of evolutionary medicine has been a total game changer when it comes to understanding best practices in the health professions. The basic principle of evolutionary medicine is pretty straightforward. The idea is that best practices in health care should always include some understanding of the evolutionary biology that underlies the particular issue at hand.

Perhaps the most conspicuous example of evolutionary medicine in action pertains to pregnancy sickness. Before people like Marge Profet (1992) applied an evolutionary lens to an understanding of the fact that many women become nauseous during pregnancy, the most common way to address pregnancy sickness was to treat it as an unpleasant symptom. So in typical Westernized medicine fashion, pharmaceuticals were created and administered. Well, if you know your medical history well, you know the drugs that were created had many adverse long-term effects on developing fetuses. This approach was disastrous.

Profet's evolutionarily informed approach to pregnancy sickness was different. It took an adaptationist approach—asking if pregnancy sickness might be some kind of biological adaptation—and, if so, we should think about the function that it addresses for the human organism. It turns out that pregnancy sickness acts exactly like an evolutionary adaptation. It is common in women across all cultures. It emerges just at the time when the basic organs are forming in the fetus. And it clearly acts to efficiently (if unpleasantly) remove potential toxins (found, i.e., in bitter vegetables) so the baby is not harmed: evolutionary adaptation—bingo.

Evolutionary Principles and Health

The entire field of evolutionary medicine looks at issues of health in this way, and it is radically improving our understanding of health issues at breakneck speed. To understand the connections between evolutionary principles and health, consider the five health-related points that follow.

1. Making People Healthy Versus Making People Feel Good

One of the core examples given regarding evolutionary medicine has to do with how to treat a fever or, in fact, how *not* to treat a fever! Sure, fever can be uncomfortable—and when it is too uncomfortable, we may need to take ibuprofen or something similar. This said, we can step back and ask why running a fever is a common response to viruses and bacterial infections. It turns out that the extreme heat associated with fever is designed to kill the infiltrating microorganisms within our bodies. So if you mask the symptom, you are not letting your body do what it needs to do to help you get better.

Optimal health (be it physical or mental) is not always associated with *feeling good*.

2. Sleep

An evolutionary perspective sheds light on all kinds of things that we take for granted. Take sleep, for instance (see Nunn, Samson, & Krystal, 2016). When examined from an evolutionary perspective, we can start to understand the function of sleep, which clearly helps with physical repair of the body by facilitating neuronal recharge in our brains. The process of sleep ultimately helps with such things as attention and other basic cognitive functions. In other words, sleep has adaptive value.

But sleep in modern Westernized people is totally out of whack. We now have entire disciplines within medicine dedicated to issues of sleep.

Part of the problem has to do with evolutionary mismatch. We have many unnatural stimuli in our lives that have adverse effects on sleep—things that did not exist during the time of human evolution thousands of generations ago. Coffee late in the day, tobacco, alcohol—all of these are modern "conveniences"—all play a role in insomnia. And the blinking blue lights of your alarm clock and cell phone are no better. Humans were meant to sleep in the dark—without having coffee with dessert.

3. Ovulation Effects

Many issues specific to women's health have been documented by this new focus on the evolution and health interface. On this point, a great deal of work has demonstrated that the female ovulatory cycle is a basic part of life as a woman. It has natural behavioral effects that relate to a variety of social outcomes (e.g., ovulating women show increased sex drive; see Geher & Kaufman, 2013).

An implication of all this relates to the large-scale use of hormonal contraceptives. Hormonal contraceptives are highly convenient and effective. This said, they are pretty unnatural at the same time. They essentially trick a woman's body into thinking that it is pregnant, thus halting the release of the egg each month. It is actually a brilliant technology—one that changes internal psychological states along with manifest behavior. A great deal of research has shown that women who do not ovulate behave differently from other women in a host of ways. This is certainly something to think about at the very least.

4. Exercise

While we are not advocating CrossFit in particular, it's noteworthy that CrossFit and various other modern exercise movements explicitly adopt an approach based on evolutionary mismatch (see Claudino et al., 2018). From an evolutionary perspective, it is clear that modern humans simply do not get the right amount of exercise. Our ancestors in the African savanna were all nomadic, and they would walk or run up to about 20 miles in a typical day. We now have "the luxury" of being able to sit at a desk all day.

From an evolutionary perspective—and from pretty much any health-related perspective—this is a problem! So we need to get ourselves to exercise. CrossFit is actually an exercise regimen that is based on an evolutionary approach, including a variety of exercises that are based on an assessment of the kinds of exercises found in pre-Westernized cultures (and that likely parallel the exercise regimens of our ancestors).

5. Diet

Perhaps the most conspicuous example of how evolutionary psychology informs health is found in an understanding of diet (see O'Keefe, Cordain, Jones, & Abuissa., 2006; Wolf, 2010). Humans have preferences for foods that are high in fat and high in sugar content because such foods were rare and were, thus, highly physiologically valuable under ancestral conditions, when drought and famine were common.

Due to modern technologies and advances in agriculture, we now have all kinds of processed foods, including foods that are unnaturally high in sugar and fat content. In fact, in a presentation given at the SUNY New Paltz Evolutionary Studies Seminar Series, Amanda Guitar (2017) pointed out that in a typical American diet, over 60% of what is eaten on a daily basis is processed foods. *Think about that.* You don't have to scratch your head too hard to see how this translates into the current obesity epidemic and resultant problems, such as relatively high rates of cardiovascular disease in Westernized (as opposed to non-Westernized) cultures. In fact, work by positive psychologists that has focused on psychological outcomes connected with nutrition has found that added sugars and a lack of nutrient-dense natural foods regularly corresponds to markers of poor mental health (see O'Neil et al., 2014; Wattick, Hagedorn, & Olfert, 2018). Our diet connects with both our physical and our mental health, and evolutionary mismatch can help us understand why.

The five topics just addressed are not thought to be comprehensive. They are, in fact, simply examples of health-related topics that have been illuminated from an evolutionary perspective.

Mental Health and Modern Technology

From the perspective of evolutionary psychology, evolutionary mismatch is often at the core of psychological problems. People become addicted to human-created technologies that exploit our evolved preferences, and addiction is associated with a whole battery of adverse psychological and physical outcomes.

Consider the following: A group in our lab (Planke & Geher, 2017) recently conducted a study on the mental health outcomes associated with being separated from family during the college years (e.g., by going to college out of state, which is, of course, very common). The findings were complex, and we won't bore you with nuances here. But we will provide you with one headline that struck us: In our sample of over 200 college students, 43.5% reported having been diagnosed at some point with some psychological disorder. Note: This is not 43.5% reporting that they *sometimes* feel anxious or depressed. This question very explicitly asked if they had been *diagnosed* with a disorder. And these findings are on a par with findings from other research teams around the country.

iGen and Mental Health Problems.   Jean Twenge (2017) discusses the iGen—or the Internet generation. This is the generation of people born between about 1995 and 2005, for whom the Internet and related technologies have always existed. This is a dramatic event in terms of human evolutionary psychology, widening the gap between modern and ancestral contexts exponentially.

As with any technological advance, Internet technology (including texting, emails, Snapchat, Instagram, etc.) was designed to improve the human condition. And, of course, in many ways, it has. For instance, there is currently, for the most part, no need for anyone in the world to own an encyclopedia, a phone book, or a dictionary.

As evolutionists, we apply an evolutionary lens to questions of behavior (see Geher, 2014). In doing so, we often ask whether modern problems are, at least partly, a result of evolutionary mismatch, which exists when some feature of our modern environment does not match the ancestral conditions that typified worlds of our ancestors during the lion's share of human evolution.

Think about iGen (also often referred to as *Generation Z*). Internet technologies have existed during their entire lives. They grew up with this stuff. They have cell phones and are, for the most part, completely and utterly addicted to them. They use such social media as Snapchat incessantly. Sure, they have the capacity to look up information, learn about all kinds of topics, and organize their lives in ways that young people before them never could. But, of course, there is no such thing as a free lunch—everything has costs.

One way that smartphones (and related technologies) may be contributing to the mental health crisis of our time is through the deindividuated communication that often occurs.

A standard finding in the social psychological literature is that people act in a relatively antisocial manner when their identities are covered up—when they are acting anonymously (see Figueredo, Vásquez, Brumbach, & Schneider, 2006; Zimbardo, 2007). This said, think about how deindividuated people are on the internet! There are all kinds of sites where people can provide anonymous feedback about businesses, doctors, professors, teachers, and more. Kids of the iGen are regularly interacting with people from all over the place remotely, so even when they are interacting with someone they know, that person is still behind a screen—literally and figuratively. Under ancestral conditions, people only would have communicated with one another in face-to-face contexts—devoid of deindividuation. In short, it's much easier for kids to be mean to one another than it ever has been before in human history—thanks to the Internet. And hurtful social outcomes can, without question, have substantive consequences regarding mental health.

Further, cell phones can be genuinely addicting. Just about everything that humans evolved to want can be found in unprecedented frequencies via our cell phones. Social interaction, social approval, sex, excitement, relationships—you name it. It's no wonder that a recent CNN poll found that 50% of teens are addicted to their cell phones. Such addictions are known to affect sleep, social interactions, and all kinds of other foundational aspects of daily life. It's important to note that *an addiction is an addiction*. Research on the nature of addiction has shown that the brain circuits associated with addiction of any kind are pretty much the same, regardless of the content of the addiction (see Montgomery, 2010). Cell phone addiction is not that different from addiction to anything else—and it is wreaking havoc on our kids' mental lives.

Addiction to Internet technologies is rampant in their generation, and there are all kinds of adverse unintended consequences that are likely to have long-term societal and personal effects on a massive scale. Do we need to suddenly cut out all social media and modern technology cold turkey? Of course that's not going to work, but it is time that we take systematic measures to address issues associated with Internet technology use by our youth. We need to do this for our kids—and we need to do this for our shared future.

Table 9.1 Human Communication Then and Now

	Preagrarian Times	Modern Times
Communication platforms	Face to Face	• Cell phones • Computers • Phones • Newspapers • Magazines • Social media
Communication partners	Kin and other individuals in one's band, usually with long-term connections among the individuals communicating with one another	• Kin • Friends • Contacts at work • Contacts related to volunteer organizations • Strangers in face-to-face interactions • Strangers in remote locations
Degree of anonymity in communication	Anonymity in communication necessarily rare	Communication under anonymous conditions very common

Table 9.1 explicates a few ways that modern forms of communication differ from ancestral forms of communication:

Healthy Goals From an Evolutionary Vantage Point

The previous sections in this chapter addressed psychological problems. However, importantly, a basic premise of positive psychology is to explore the other side of our psychological worlds—the positives, the things that we are doing right. This section is dedicated to exploring positive psychological processes that facilitate mental health via an evolutionary lens.

The evolutionary perspective has many implications for the pursuit of happiness and well-being. These implications address a broad range of issues, including stress reduction, weight management, and managing addictions.

Practice Mindfulness

Under modern Westernized conditions, we are bombarded with stimuli of all kinds. Many of these stimuli either are unnatural or are presented in doses that are considerably higher than would have been the case under ancestral conditions. The cacophony of stimuli that we experience on a daily basis is completely disproportionate from what would have been experienced by our ancestors prior to agriculture and industrialization. For this reason, efforts to seek mental peace are essential in this day and age. In other words, many modern movements to facilitate mental or spiritual hygiene can be thought of as ways to reduce evolutionary mismatch related to stimulus overload in modern contexts. One large focus of modern psychological research that takes this approach deals with *mindfulness*.

The practice of mindfulness has gained a lot of popularity lately. Mindfulness training has been shown to reduce stress (de Vibe et al., 2013); increase and enhance well-being (O'Leary & Dockray, 2015); reduce anxiety (Mayorga, De Vries, & Wardle, 2016); contribute to weight management (Van de Veer, Van Herpen, & Van Trijp, 2016); decrease intense pain (Anheyer et al., 2017); decrease alcohol and drug use (Wupperman et al., 2015); and even contribute to increased relationship and sexual satisfaction (Khaddouma, Gordon, & Bolden, 2014).

Mindfulness is a state of consciousness (see de Vibe et al., 2013). It's an awareness of what is going on in the present—a peaceful appreciation of the here and now. It is often practiced through gentle meditation and breathing techniques. Mindfulness practices typically help people to slow down, focus on their breathing, and begin to simply notice—not change or judge—just notice the sounds around them, their physiological state, their mental state, and their environment. There's a whole world of resources available to us these days for starting to practice mindfulness techniques; classes, books, and apps on your smartphone are all easily accessible for those looking to see how mindfulness might increase well-being.

One particularly unique mindful human experience pertains to the experience of awe (see Yaden et al., 2018). Imagine the sensation you get when you listen to music that gives you the chills or the sensation you get when you realize how tiny we are in the vast greatness of this universe. Awe is part of a mindful experience; it's a complex emotion that has been frequently associated with a sense of openness and other positive experiences. Getting lost in art or recognizing beauty, virtue, or even something supernatural can all contribute to sensations of awe, which we may think of as mindfulness at its best.

Cultivate Friendships

Something else we can dedicate our time to in an effort to increase and strengthen our healthy mindset is friendship. This may sound obvious, but making time for loved ones is a serious boost.

As described in prior sections of this book, for humans, reciprocal altruism is a basic feature of our social ecologies (R. Trivers, 1985). In such a world, developing genuine friendships (independent of kin) is an essential part of our evolutionary heritage. Don't blow it! People evolved to help nonkin—with expectations of being helped in return—and we evolved to have expectations of such relationships between reciprocating individuals lasting for a long time. So be a loyal friend, like the most successful of our ancestors surely were.

Expect a Long Social Life

In some species, such as beavers, an adult animal can go months without seeing a *conspecific* (a member of its same species). In other species, such as North American crows, animals see the same individuals day in and day out, across seasons and years. Humans are more like crows than like beavers. In such species, animals form long-term relationships with several specific individuals. They come to rely on one another for help in such tasks as finding and sharing food. What's good for the goose is good for the gander— regardless of kin lines, in many cases. Humans are perhaps the world's leading prototype of a species that has a consistent social group across long periods of time. Let this fact help guide your interactions, and you'll benefit accordingly.

Treat Others Like You Live in a World of 150 People

Under modern conditions, we are often surrounded by strangers we've never seen before and will likely never see again (think of being on a train in a foreign country). Under ancestral conditions that typified hominid evolution for thousands and thousands of generations, humans rarely encountered *any* individuals outside their own clan. These clans were stable groups, including both kin and individuals with long-standing relationships with clan members, typically totaling about 150 individuals (Dunbar, 1992). If you were only going to see the same 150 people—and only those 150—for the next 40 or so years, how would you treat them? Kindly, of course!

So, how does this all relate to a healthy mind? We know that friendship, love, and connectedness to those around us have always been a part of our lifestyle. But what do the data say about it? In short, studies across a broad array of platforms have shown that maintaining a loving and connected circle of relationships contributes to greater happiness and a better memory into later age when compared to more isolated individuals (Waldinger, 2019). In short, research into the effects of relationship quality and life satisfaction has found that good relationships with others are what keep us happier and healthier. Other variables don't generally matter as much; maintaining meaningful relationships with others is a cornerstone of what keeps us happiest and healthiest. Turns out, this part of our ancestral past has been highly adaptive throughout our evolutionary history and remains so today.

Appreciate Nature

When astronauts spend time in outer space, they often focus energy on life forms from Earth—such as the wheat cultivated on the Russian space station *Mir* by American astronaut John Blaha, and others, in the 1990s. One might wonder, of all the things to focus one's energy on in outer space, why would someone choose to spend hours a day watching wheat grow?

The field of evolutionary psychology gives us good insights into the *wheat is interesting in space* phenomenon. Evolutionary psychology is largely premised on the idea that the human mind includes a multitude of adaptations and psychological processes that evolved before the rise of large-scale civilization. As addressed by various examples in this chapter, it turns out that in modern contexts, we often find environments to be out of synch with our mental proclivities.

While spending time on a space station with some Russian guys sounds pretty cool, it is also evolutionarily unnatural—such a situation provides all kinds of stimuli and situations that would have never been encountered by our ancestors. From this perspective, it's little wonder that astronauts will spend hours a day watching plants grow as plant life has been part of all hominid environments—forever. And plants provide significant resources that humans have relied on for our entire existence.

Based on the idea that humans evolved in out-of-door environments for thousands of generations, you would expect that humans have a natural inclination toward things found in nature—items that would have some bearing on survival. And, in fact, research shows that this is exactly the case (see McMahan, Cloud, Josh, & Scott, 2016). Atran (1998) has strongly documented that human cognitive processes are honed for nature, with people all across the world having dedicated psychology for categorizing plants versus animals, for instance.

Along these lines, there's strong evidence that people are particularly attracted to natural environments that typify the African savanna that our ancestors evolved in (Orians & Heerwagen, 1992). We like to look at trees, animals, and water; of course, all these things had important implications for the survival and ultimate reproduction of our ancestors, so it makes good sense that we would have evolved to pay attention to these environmental features.

In a broad sense, the great evolutionary biologist E. O. Wilson (1984) used the term *biophilia*—the love of living things—to characterize the human mind, and for the money, this sounds spot on. Next time you just want to jump in a lake on a summer day, hike through the woods up a rocky mountain, or walk along the tidal coast in wonder at the nature that surrounds you, realize that you're not alone—our love of the out-of-doors is foundational to human evolutionary psychology.

People are physically active for a lot of reasons—to stay physically fit, for physiological health reasons, to alleviate anxiety, etc. A great way to stay physically active is to get outside and move. It's natural, after all. And it might keep you away from looking at your phone every 2 minutes!

If you're looking to increase your physical activity but don't know where to start, there's no need to sign up for a marathon just yet. Start by taking walks around the neighborhood after dinner or go throw a ball in the park. Getting outside and moving is as much a part of our heritage as is eating, sleeping, and sex. It's something that we should remember to make time for.

Implications for Positive Psychology

Positive psychology centers largely around mental, physical, and communal health. By taking an evolutionary approach to studying these human capacities, we begin to see that practices like selflessness, virtuosity, temperance, altruism, and even diet and exercise are nothing new. There's no need to reinvent the wheel here per se. Instead, we need to look back and remind ourselves of how the wheel has already been for millions of years. By studying these facets of positive psychology through an evolutionary lens, we can see more clearly how these positive emotions and practices of healthy living have been adaptive for humans all along. Then we can begin to practice them more regularly and with more intention if we wish to live healthfully and truly flourish.

Bottom Line

So, let's be more cautious and aware of our smartphone habits. Let's put the phone away now and focus on face-to-face interactions and quality time with those who matter to us. Let's take time to foster and nourish meaningful relationships with others. And *do* give mindfulness practice a shot! You may find it to be helpful as it adds a richness and calmness to your life. But don't attempt to eliminate any and all less favorable emotional states as these are equally as adaptive as happiness. A healthier lifestyle when it comes to the mind must include a range of human emotions and qualities—not just the comfy ones.

So make time each day to go outside and move around. Explore nature and get your heart rate up every now and again. Remember to cultivate social relationships with kindness and respect. And take some time out of mind each day to reduce the evolutionarily unnatural stimulus overload that is entrenched in so many modern environments. Whether you do that through meditation, yoga, the practice of selfless love, paying it forward, or any other means, let's remember to feed our inner selves with some peace and some purpose.

Acknowledgments

Some content from this chapter was adapted from Glenn Geher's (2014) *Psychology Today* blog post "10 Ancient Rules We Should All Live by Today," his (2013) post "Why We Love the Out of Doors," and his (2017) post "The Mental Health Crisis Is Upon the Internet Generation." Glenn owns the copyright to the material. Some of the content comes from Glenn's invited

Evolution Institute post "The Rise of Deindividuated Communication: An Evolutionary Perspective on the Real Problem With Increased Screen Time." This content is being used with the permission of the Evolution Institute.

References

Anheyer, D., Haller, H., Barth, J., Lauche, R., Dobos, G., & Cramer, H. (2017). Mindfulness-based stress reduction for treating low back pain: A systematic review and meta-analysis. *Annals of Internal Medicine, 166,* 799–807. doi:10.7326/M16-1997

Atran, S. (1998). Folk biology and the anthropology of science: Cognitive universals and cultural particulars. *Behavioral and Brain Sciences, 21,* 547–609.

Claudino, J. G., Gabbett, T. J., Bourgeois, F., Souza, H. de S., Miranda, R. C., Mezêncio, B., . . . Serrão, J. C. (2018). CrossFit overview: Systematic review and meta-analysis. *Sports Medicine—Open, 4,* 11.

De Vibe, M., Solhaug, I., Tyssen, R., Friborg, O., Rosenvinge, J. H., Sorlie, T., & Bjorndal, A. (2013). Mindfulness training for stress management: A randomised controlled study of medical and psychology students. *BMC Medical Education, 13,* 107. doi:10.1186/1472-6920-13-107

Diener, E., Fraser, S. C., Beaman, A. L., & Kelem, R. T. (1976). Effects of deindividuation variables on stealing among Halloween trick-or-treaters. *Journal of Personality and Social Psychology, 33,* 178–183.

Dunbar, R. I. M. (1992). Neocortex size as a constraint on group size in primates. *Journal of Human Evolution, 22,* 469–493.

Figueredo, A. J., Vásquez, G., Brumbach, B. H., & Schneider, S. M. R. (2006). The heritability of life history strategy: The K-factor, covitality, and personality. *Social Biology, 51,* 121–143.

Fisher, H. (1993). *Anatomy of love—A natural history of mating, marriage, and why we stray.* New York: Ballantine Books.

Geher, G. (2013). Why We Love the Out of Doors. *Psychology Today* blog post.

Geher, G. (2014). *Evolutionary psychology 101.* New York: Springer.

Geher, G. (2014). 10 Ancient Rules We Should All Live By Today. *Psychology Today* blog post.

Geher, G. (2017). The Mental Health Crisis is upon the Internet Generation. *Psychology Today* blog post.

Geher, G. (in process). The Rise of Deindividuated Communication: An Evolutionary Perspective on the Real Problem With Increased Screen Time. Invited blog post for the *Evolution Institute.*

Guitar, A. E. (2017). *Evolutionary medicine.* Presentation given for the Evolutionary Studies Seminar Series, New Paltz, New York.

Khaddouma, A., Gordon, K. C., & Bolden, J. (2015). Zen and the art of sex: Examining associations among mindfulness, sexual satisfaction, and relationship satisfaction in dating relationships. *Sexual & Relationship Therapy, 30*(2), 268–285. doi:10.1080/14681994.2014.992408

Liechty, J. M., & Lee, M. (2013). Longitudinal predictors of dieting and disordered eating among young adults in the US. *International Journal of Eating Disorders, 46,* 790–800. doi:10.1002/eat.22174

Mayorga, M., De Vries, S., & Wardle, E. A. (2016). Mindfulness behavior and its effects on anxiety. *Journal on Educational Psychology, 9*, 1–7.

McMahan, E. A., Cloud, J. M., Josh, P., & Scott, M. (2016). Nature with a human touch: Knowledge of human-induced alteration impacts perceived naturalness and preferences for natural environments. *Ecopsychology, 8*, 54–63.

Montgomery, J. (2010). *The answer model: A new path to healing.* TAM Books.

Nesse, R. M., & Williams, G. C. (1995). *Why we get sick: The new science of Darwinian medicine.* New York: Times Books.

Nunn, C. L., Samson, D. R., & Krystal, A. D. (2016). Shining evolutionary light on human sleep and sleep disorders. *Evolution, Medicine, and Public Health, 1*, 227–243.

O'Keefe, J. H., Cordain, L., Jones, P. G., & Abuissa, H. (2006). Coronary artery disease prognosis and C-reactive protein levels improve in proportion to percent lowering of low-density lipoprotein. *The American Journal of Cardiology, 98*, 135–139.

O'Leary, K., & Dockray, S. (2015). The effects of two novel gratitude and mindfulness interventions on well-being. *The Journal of Alternative and Complementary Medicine, 21*, 243–245.

O'Neil, A., Quirk, S. E., Housden, S., Brennan, S. L., Williams, L. J., Pasco, J. A., . . . Jacka, F. N. (2014). Relationship between diet and mental health in children and adolescents: A systematic review. *American Journal of Public Health, 104*, e31–e42. http://doi.org/10.2105/AJPH.2014.302110

Orians, G. H., & Heerwagen, J. H. (1992). *Evolved responses to landscapes.* In J. Barkow, L. Cosmides, & J. Tooby (Eds.), *The adapted mind: Evolutionary psychology and the generation of culture* (pp. 555–579). New York: Oxford University Press.

Planke, J., & Geher, G. (2017). *Evolutionary mismatch and the modern American family: Implications for mental health.* Presentation given at annual meeting of the NorthEastern Evolutionary Psychology Society, Binghamton, NY.

Profet, M. (1992). Pregnancy sickness as adaptation: A deterrent to maternal ingestion of teratogens. In J. Barkow, L. Cosmides, & J. Tooby (Eds.), *The adapted mind: Evolutionary psychology and the generation of culture* (pp. 327–365). New York: Oxford University Press.

Trivers, R. (1985). *Social evolution.* Menlo Park, CA: Benjamin/Cummings.

Trivers, R. L. (1971). The evolution of reciprocal altruism. *Quarterly Review of Biology, 46*, 35–57.

Twenge, J. (2017). *Why today's super-connected kids are growing up less rebellious, more tolerant, less happy—And completely unprepared for adulthood—And what that means for the rest of us.* New York: Simon and Schuster.

Van De Veer, E., Van Herpen, E., & Van Trijp, H. C. M. (2016). Body and mind: Mindfulness helps consumers to compensate for prior food intake by enhancing the responsiveness to physiological cues. *Journal of Consumer Research, 42*(5), 783–803.

Waldinger, R. (2019). The pursuit of happiness. *In progress.*

Watkins, P. (2014). *Positive psychology 101.* New York: Springer.

Wattick, R. A., Hagedorn, R. L., & Olfert, M. D. (2018). Relationship between diet and mental health in a young adult Appalachian college population. *Nutrients, 10*, 957. doi:10.3390/nu10080957

Wegner, D. M. (2011). Setting free the bears: Escape from thought suppression. *American Psychologist, 66*(8), 671–680.

Wilson, D. S. (2002). *Darwin's cathedral: Evolution, religion and the nature of society.* Chicago: University of Chicago Press.

Wilson, E. O. (1984). *Biophilia*. Cambridge, MA: Harvard University Press.

Wolf, R. (2010). *The paleo solution*. Las Vegas, NV. Victory Belt.

Wupperman, P., Cohen, M. G., Haller, D. L., Flom, P., Litt, L. C., & Rounsaville, B. J. (2015). Mindfulness and modification therapy for behavioral dysregulation: A comparison trial focused on substance use and aggression. *Journal Of Clinical Psychology, 71*, 964. doi:10.1002/jclp.22213

Yaden, D. B., Kaufman, S. B., Hyde, E., Chirico, A., Gaggioli, A., Zhang, J. W., & Keltner, D. (2018). The development of the Awe Experience Scale (AWE-S): A multifactorial measure for a complex emotion. *The Journal of Positive Psychology*. https://psycnet.apa.org/doi/10.1080/17439760.2018.1484940

Zimbardo, P. (2007). The Lucifer Effect: Understanding How Good People Turn Evil. *The Journal of the American Medical Association, 298*, 1338–1340.

Zunker, C., Peterson, C., Crosby, R., Cao, L., Engel, S., Mitchell, J., & Wonderlich, S. (2011). Ecological momentary assessment of bulimia nervosa: Does dietary restriction predict binge eating? *Behaviour Research and Therapy, 49*, 714–717.

10

Building Darwin's Community

The positive psychology movement is intended to be just as focused on the flourishing of communities as it is on the flourishing of individuals (see Seligman, Steen, Park, & Peterson, 2005). This chapter looks outward and, as such, focuses on applications of evolutionary psychological principles to questions of communal functioning.

The Evolutionary Psychology of Community

In many ways, humans are the communal ape. There is no escaping this fact. As discussed in Chapter 4, focused on politics, at some point in our evolutionary history, our ancestors formed groups, or communities, that extended beyond kin lines (see Bingham & Souza, 2009; Wilson, 2007). Once this happened, all kinds of possibilities emerged. Communities of humans, working together, were able to achieve all kinds of great outcomes as a result.

While there are lots of ways to think about how humans are unique or distinct from members of other species, our tendency to form communities beyond kin lines definitely stands out as a significant one. To put a man on the moon, you need more than just a group of smart individuals, a family, or even a group of smart families. Even a village isn't quite enough. You need a community. And much of the success of *Homo sapiens* resides in the fact that we are, in a palpable way, the community-building ape.

In creating a community of any sort, we create a very legitimate psychologically based entity that allows us to be, in an important sense, part of something bigger than ourselves.

What Is a Community? Communities are, to a large extent, psychologically and socially constructed. To put a face to the idea of community, think about my (G. G.'s) life. I belong to several communities. I see my little neighborhood in upstate New York as a community. I'm a member of the faculty at SUNY New Paltz—that's a community. I'm a member of several groups of scholars and students who are passionate about the ways that evolution has influenced human behavior. I'm a member of the Hudson Valley disk golf community. I'm a member of the Hudson Valley running community. I'm in an all-professor punk rock band, Questionable Authorities, and this is kind of its own community. And there is more. In our modern, often-busy lives, we have choices to make. When you stop to think about the many different communities that you belong to at any given time, you'll quickly appreciate the fact that we are communal by nature.

To get a solid sense of what a human community is like—and how we can better understand the concept of community via concepts from evolutionary psychology—let's consider a particular example. In the little corner of the world where my family is fortunate to live, the Hudson Valley of New York, there are all kinds of opportunities to be part of and cultivate community life. An amazing example is found in the group formed by Hudson Valley icon Erica Chase-Salerno, called, simply, Hudson Valley Parents. When my wife, Kathy, and I (G. G.) moved to the Hudson Valley in 2000, we knew few people. And we were already in the process of starting a family. When our daughter, Megan, was born, Kathy looked into parent groups, and she found the then-fledgling *Mommy's Group*, which soon morphed into the more inclusive *New Paltz Playgroup*. This was an informal organization for parents and families in the area to connect and share resources and time. Early in the process, Erica, mother of two kids, Declan and Quinn, along with her helpful and tech-savvy husband, Michael, joined in. And, seeing many benefits to this organization for many people, she took it on herself to expand this group.

A natural leader with a reputation for helping others, Erica developed a listserv, website, and Facebook group for this community, which now serves thousands of families on a regular basis. Erica's work in developing the Hudson Valley Parents Network has been exemplary in terms of what it means to build and cultivate a community. This organization connects newly transplanted parents to a solid group of caring others, concurrently providing much-needed informational and communication-based resources for thousands of families across a broad geographical region.

While this story partly tells a tale about an amazing woman who goes the extra mile to help with the greater good, it also tells a tale about human potential when it comes to community building. Building communities that cut across lines of kinship—creating "families" of individuals who share goals rather than genes—is an important and positive part of our evolutionary heritage.

Neighborhoods as Communities

It turns out that an evolutionary perspective on the nature of community can lead to all kinds of novel research questions and findings. For instance, in research on human social communities from an evolutionary perspective, David Sloan Wilson and Dan O'Brien (O'Brien & Wilson, 2011; Wilson, 2011) examined behaviors of individuals within neighborhood communities from an evolutionary–ecological perspective. That pretty much means that we can understand neighborhood behaviors by applying an evolutionary lens—asking how patterns of behaviors of individuals within neighborhoods serve the individual actors, their families, their neighborhoods, and their broader communities.

Social Capital

One thing that Wilson, O'Brien, and their colleagues focus on in regard to neighbor-related behavior pertains to social capital, a term that generally pertains to how well the people in a neighborhood demonstrate respect for the neighborhood and display markers of connections to one another. You can see how neighborhoods vary on this one when it comes to Christmas decorations. In some neighborhoods, people go over the top—white lights, colored lights, blinking lights, Santa, Snowman, blow-up Grinch, etc. In other neighborhoods, you'd barely know what season it is. Neighborhoods that get all decked out for the holidays can be thought of as higher in social capital, and it turns out that when you study people from these neighborhoods compared with neighborhoods that are lower in social capital, the folks from neighborhoods with high social capital feel safer and more connected to others in their community. What is the function of getting out the Christmas decorations each year? Well it may be a lot of work, but it's a great way to display social capital, which is a glue that connects people to one another.

Neighborhoods in Disrepair

On the flip side, neighborhoods that are in disrepair don't "look as nice." They may be laden with potholes, weeds coming out of the sidewalks, broken streetlights, and the like. You might wonder how these seemingly aesthetic details actually affect quality of life. It turns out that they do. In neighborhoods that suffer from disrepair—that is, neighborhoods low in social capital—people behave less altruistically toward one another. They feel less connected to one another, and they feel less safe. You have to think, if you're looking to rob a house (please don't!), you would probably see a house in a neighborhood with low social capital as an easier opportunity compared with one with high social capital because all the features of a neighborhood with high social capital signal, "We are here, we care about this place, and we keep an eye on one another. Beat it!" A straight-up stop sign at the end of the road says this. A stop sign that sits cockeyed for months on end says quite the opposite.

Interestingly, these effects of social capital that O'Brien, Wilson, and their colleagues documented existed largely independently of socioeconomic status. So among two very wealthy neighborhoods, the one with higher social capital tends to have people who feel safer and better connected to others— and among two very poor neighborhoods, the same pattern applies. Displays of social capital seem to be a crucial way that modern humans display territory and provide security for those close to them.

Custodians of the Neighborhood

In a great set of studies conducted under the auspices of the Boston Area Research Initiative, O'Brien, Gordon, and Philippi-Baldwin (2014) examined closely the tendency for individuals to act as *custodians* within their neighborhoods. This research, conducted throughout the city of Boston, used calls to the city's 311 hotline as a marker of prosocial behavior. The 311 hotlines, becoming common among cities all around, is sort of like 911 light. It's used for calling in things that should be brought to the attention of the municipality but are not life threatening. For instance, you'd call 311 to report graffiti or a broken streetlight.

By using GPS technology, O'Brien and his team have been able to map patterns of 311 reporting behavior. And the data tell some very interesting stories—stories that betray our evolutionarily based mindset. Two of O'Brien et al.'s main findings are the following:

1. **Few people take on a custodial role within a neighborhood.** Defining custodians as those who make multiple calls to 311 within a specified

time period, it looks like only about 3–6% of the folks in a neighbor-hood seem to take on this role.

2. **Custodial behavior tends to be neighborhood specific.** When you look at the patterns of calls from people who make multiple 311 calls, it's clear that people are *way* more likely to call about problems in their imme-diate neighborhoods or nearby (within a few blocks). If Joe lives in the North End and finds himself in Chinatown one day and he spots some graffiti down there, he's less likely to report that compared with the graf-fiti that he finds on the stop sign across the street from his apartment when he gets back home. Prosociality has geographical boundaries.

When you step back and examine custodial behavior, there's clearly a self-interested element to it.

Those of us who take actions to look after our communities are playing out ancestral themes when it comes to human social behavior. What's good for one's community is, ultimately and importantly, good for oneself and this person's broader circle.

Why Do We Mow the Lawn, Rake the Leaves, and Plant Nice Bushes?

If you're like many Americans, then a good percentage of your time across the year is spent mowing the lawn, raking the leaves, shoveling the driveway, setting up Christmas lights, taking down Christmas lights, etc. Your house might not make you the Joneses of your neighborhood, but you probably make sure to keep things up to par as best as you can. Why do we do this? Why do we take care of our properties and, by extension, our neighborhoods and communities therein? The answer partly is this: Doing so displays social capital—a glue that interconnects people and that creates perceptions of se-curity. Have we evolved so we try to establish harmonious social connections and a sense of security for our family and friends? Yeah, we think so!

Some Takeaways From the Communal Ape

This section includes some takeaways for life that follow from the fact that humans are a communal ape.

The More You Give, the More You Will Receive

Want to do something great in life? In a communal species such as ours, perhaps start by cultivating a reputation as a giver. Research shows that *reciprocal altruism*, or the tendency of individuals to exchange goods, is a fundamental human trait. We are an altruistic organism—when you help someone, there's an implicit understanding that help will come to you in return. Through this basic mechanism, large-scale social groups, communities, and alliances can emerge. Given that this is how our species operates, developing a reputation as a giver is essential; it signals to others that you would be a good target for their altruistic acts (as they could confidently expect payback at some future point from a giver like yourself).

Acts of Kindness Make Everyone Feel Good

Given how evolutionarily beneficial it is to give to others, it's no wonder that our proximate psychology helps reinforce giving behavior. Altruistic acts inherently benefit both the recipient and the altruist: Much research (e.g., Underwood, Froming, & Moore, 1977) has found that giving *feels good*. One receives the good given, and the giver feels an emotional high.

In the way of anecdotes, I (N. W.) can tell you that I am personally a more energetic and happy person when I'm involved in my community. I wake up with more pep in my step when I have a purpose that may benefit others.

I've moved around a lot. Young as I may be, I've lived in several different states. I was born and raised in Southern California. After I completed my undergraduate degree, I packed up my life and my cat into my '02 Honda Civic and drove to upstate New York to study evolutionary psychology. After I completed my graduate degree, I moved to Denver, Colorado—really just for the heck of it. Now, I'm happily living with both feet on the ground in Richmond, Virginia. Here's what I've learned from all of my shuffling around: I'm happier when I put my time and energy into the community. My first big move, the one to upstate New York, was one of the scariest things I've ever done. My plan was to conduct research with the highly renowned evolutionary psychologist Gordon Gallup. Working with him, by the way, ended up being one of the best things I've ever done.

Anyway, I had no family or friends there—only my ambition and hopes. And it remained scary and lonely for a long time because I didn't know what to do or how to "get out there." I went to lab meetings to do volunteer research, I'd go to work at the bar, and I'd come home. I'd always been involved in extracurricular sports, school, exercise groups, and even a band when I was in California. Things come more naturally for people I think in their hometown or state. But you quickly realize when you throw yourself into an entirely new set of waters that you need to make a proactive effort to get out and involved in things. Over time, I became more involved in the community—I started volunteering, I made an effort to stay late after work and make friends, I found groups to join and meet up with, and my happiness and sense of well-being returned with a lot of excitement. That experience taught me so much. When I moved to my current city of Richmond, I wasted no time. Immediately, I introduced myself to my neighbors and made fast friends with them. I looked into local events to attend. I began volunteering for a group that matters to me. It's all made this experience absolutely wonderful, and I don't attribute that to my personality or individual differences. I attribute this positive outcome in my own life as following from the fact that I have come to truly appreciate the importance of community. I genuinely believe that forcing ourselves to get out in the community, however uncomfortable it may be to start, is one of the best ways to flourish in life and to feel driven. It's a way to give purpose and meaning to life.

The Nuts and Bolts of Evolved Human Communities

The theme of community runs throughout this book—for good reasons. Ancestral *Homo sapiens* differed from other Hominids largely because they had the proclivity to form communities beyond kin lines—communities with shared goals and coordinated groups of individuals. Based on the extensive literature on the topic of the evolution of human communities, the following is a list of nuts and bolts that, from an evolutionary perspective, make for healthy and successful human communities:

- **A Shared Mission**—When individuals have a shared set of goals, they work together as working toward each goal concurrently benefits both the individual and the community writ large (Wilson, 2007).
- **Norms of Reciprocal Altruism**—As described in previous sections, the fact of reciprocal altruism lies at the core of our communal tendencies.

When people help others and return help that they have received, communities tend to thrive (see Trivers, 1971).

- **Punishment of Cheaters**—There is no free lunch when it comes to being human. Based on our long history of reciprocal altruism, humans are sensitive to others who cheat by getting more than their fair share. We evolved to detect this kind of nonsense, and our best communities have systems in place to do just that (see Ermer, Cosmides, & Tooby, 2007).

- **Evolutionarily Appropriate Conditions**—As addressed extensively in previous sections of this book, humans did not evolve to live in modernized large-scale conditions. Human communities evolved under small-scale ancestral conditions. As such, leaders of modern-day communities should strongly understand the ecological conditions that our minds evolved to experience. That means that communities should approximate small-scale groups (vis-à-vis Dunbar's number) and should underscore kin relations (as do many fraternal groups, i.e., that conceptualize members as *brothers* or *sisters*). Using features of ancestral social conditions to shape modern human communities will only help modern communities thrive by capitalizing our evolved psychology.

Implications for Positive Psychology

One of the core goals of positive psychology is to help cultivate positive human communities. Based on the work in the field of evolutionary psychology, it is clear that understanding the fact that kindness, altruism, democracy, and prosocial behavior are all evolutionarily adaptive characteristics of the human condition is a great foundation for positive psychology. Understanding the evolutionary origins of these features surely can help us come up with more powerful ways to build thriving human communities.

Bottom Line

Human beings are a communal ape sine qua non. So if you want to figure out how to live the rich life, perhaps focus on how you can connect with others as part of a community. Don't isolate yourself. Being proactive in your community benefits those close to you, your neighborhood and community, and you yourself. We have a deep-rooted history that shows that kindness and altruism are really the way to go as these attributes ultimately have significant communal benefits.

Acknowledgments

Some content from this chapter was adapted from Glenn Geher's (2015) *Psychology Today* blog post "Custodians of the Neighborhood," as well as his (2015) post "The One Graduation Message We All Need to Hear." Glenn owns the copyright to the material.

Note that it is with deep regret that we write, here, that since this chapter was originally written, Erica Chase-Salerno, used as an exemplar of community herein, passed away as a result of terminal cancer. Our hearts go out to her and her family. And this chapter is dedicated to Erica, whose generosity and sense of community knew no boundaries.

References

Bingham, P. M., & Souza, J. (2009). *Death from a distance and the birth of a humane universe.* Lexington, KY: BookSurge.

Ermer, E., Cosmides, L., & Tooby, J. (2007). Cheater detection mechanism. In R. F. Baumiester & K. D. Vohs (Eds.), *Encyclopedia of social psychology* (pp. 138–140). Thousand Oaks, CA: Sage.

Geher, G. (2015). The one graduation message we all need to hear. *Psychology Today* blog.

Geher, G. (2015). Custodians of the neighborhood. *Psychology Today* blog.

O'Brien, D. T. (2015). *311 hotlines and the maintenance of the urban commons: Examining the intersection of policy and the evolved human animal.* Presentation given for the Evolutionary Studies Seminar Series, New Paltz, NY.

O'Brien, D. T., Gordon, E., & Philippi-Baldwin, J. (2014). Territoriality, attachment to space and community, and maintenance of the public space: A field study integrating administrative records of reports of public issues with self-reports. *Journal of Environmental Psychology.*

O'Brien, D. T., & Wilson, D. S. (2011). Community perception: The ability to assess the safety of unfamiliar neighborhoods and respond adaptively. *Journal of Personality and Social Psychology, 100,* 606–620.

Seligman, M. E. P., Steen, T. A., Park, N., & Peterson, C. (2005). Positive psychology in progress. Empirical validation of interventions. *American Psychologist, 60,* 410–421.

Trivers, R. L. (1971). The evolution of reciprocal altruism. *Quarterly Review of Biology, 46,* 35–57.

Underwood, B., Froming, W. J., & Moore, B. S. (1977). Mood, attention, and altruism: A search for mediating variables. *Developmental Psychology, 13,* 541–542.

Wilson, D. S. (2007). *Evolution for everyone: How Darwin's theory can change the way we think about our lives.* New York: Delacorte Press.

Wilson, D. S. (2011). The Neighborhood Project: Using evolution to improve my city, one block at a time. New York: Little, Brown.

SECTION IV

IMPLICATIONS AND THE FUTURE OF POSITIVE EVOLUTIONARY PSYCHOLOGY

This section, comprising a single final chapter, addresses implications of the field of positive evolutionary psychology for living the good life—along with a demarcation of directions for future researchers who are interested in these ideas.

In terms of future research areas, the content of this book points to a broad array of topics, such as the evolutionary psychology of religion, the evolved function of happiness, the importance of reciprocal altruism in human social interactions, and more. Part of this final chapter is dedicated to inspiring researchers who are interested in positive evolutionary psychology by providing an array of possible research areas and questions that future work can address.

This book also has something of a self-help quality to it, and this final chapter delineates several specific implications for living that one might glean based on this work. These implications bear on such issues as religion, love, and social life. All in all, this final chapter is designed to point the reader toward ways to use Darwin's big ideas to live the good life.

11

Darwin's Quick Tips for Living
a Richer Life

Positive Evolutionary Psychology Reconsidered

Every now and again, great ideas that have strong potential with other great ideas progress independently. Until Harvard biologist Ernst Mayr made the connection between Darwin's ideas on natural selection and the nature of DNA, work in evolutionary biology and work on genetics progressed independently from one another (see Mayr & Provine, 1980). When such great scholars as Leda Cosmides, John Tooby, Steven Pinker, and David Buss connected modern empirical human psychology with evolutionary biology, the field of evolutionary psychology was born. The modern field of applied behavior analysis owes to the marriage of behaviorism and work in the field of special education. Reese's Peanut Butter Cups merge peanut butter and chocolate and so forth. Sometimes, stepping back and thinking in new ways about the interconnection among existing ideas is a great way to advance our understanding of the world and our place in it.

This book outlines such an intellectual marriage. Evolutionary psychology and positive psychology are two of the most cutting-edge fields in the behavioral sciences. Each is the subject of multiple textbooks—and each is taught as a content course in college classrooms around the world. Further, to our minds, the goals of these two areas of inquiry are fully compatible with one another. As a basic area of intellectual inquiry, evolutionary psychology seeks to help advance our understanding of behavior by applying evolutionary principles. Positive psychology has a primary goal that is a bit more applied in nature. The basic goal of positive psychology is to help elucidate the positive aspects of the human psychological experience, including understanding factors that increase well-being at the individual and community levels.

As an area of scientific inquiry, evolutionary psychology has been famously effective and powerful in helping to shed light on such important human domains as physical health, psychological health, education, politics, and intimate relationships—among others. It only makes sense that the evolutionary approach in psychology would prove so productive given how famously insightful Darwin's ideas related to the nature of life are.

The basic idea of positive evolutionary psychology is simply the application of the evolutionary psychological approach to questions that bear on the field of positive psychology—how the evolutionary approach can help people to *lead the good life*, so to speak. Throughout the pages in this book, we have outlined various ways that the evolutionary psychological approach can inform the good life on which positive psychologists focus their efforts.

Future Research Directions

While this chapter is primarily about takeaway messages that positive evolutionary psychology has for anyone, we think it's important to also include some guidance for researchers who are interested in conducting work that fits with the ideas of this field. While several topics make sense as being examined in terms of the positive evolutionary psychology approach, the topics included in this section jump out to our minds as areas of inquiry that will likely emerge as particularly fruitful.

The Evolutionary Psychology of the Moral Emotions

The research that our lab has conducted on the moral emotions and how people respond to transgressions has been a truly rewarding scholarly experience. Framing questions related to the moral emotions and social outcomes from an evolutionary psychological perspective helps to very much focus the nature of research on these topics. Starting with the idea of evolutionary mismatch and Dunbar's number and integrating ideas related to reciprocal altruism into the mix, this research asks important questions about human morality, emotions, and social outcomes—all of which connect with the field of positive psychology.

Future research can delve further into such questions as the following: (a) How do reciprocated altruistic acts affect the emotional states of the altruist and the target of the altruist? (b) How do moral emotions such as guilt, shame, and forgiveness play out in a cross-cultural context? (c) How does effective decision-making in social contexts relate to such evolutionarily relevant factors as the tendency to reciprocate altruism?

The Evolutionary Psychology of Happiness and Well-Being

Positive psychologists are sometimes criticized for focusing too much on factors associated with happiness—in spite of relevant contextual factors. The positive evolutionary psychological perspective has much to contribute in this regard. Adding evolutionary concepts into the mix allows for an assessment of happiness and well-being in a broader context. Thus, an evolutionary psychologist will not simply ask how we can make people happier—an evolutionary psychologist will first consider the evolutionary origins of emotions such as happiness and will then think about how advancing human happiness now connects with the evolutionary function of happiness. Before starting an all-out program to increase human happiness, the evolutionist asks if unlimited happiness makes sense as a goal given that all of the negative emotions, such as fear and anxiety, evolved for good evolutionary reasons. The evolutionary perspective gives an important and biologically derived framework for thinking about the human emotions and human well-being.

Research on this topic can examine such questions as (a) What factors are universally associated with happiness in humans? (b) What are the physical, social, and psychological effects of increasing positive affect to potentially abnormal levels? (c) Is there a natural or even optimal ratio of positive and negative emotions to one another that can be understood in terms of human evolution?

The Evolutionary Psychology of Religion and Spirituality

Thanks to the great work of scholars such as David Sloan Wilson (2002), the conversation regarding the evolution and religion interface has very much gotten past the caricature of this conversation, which focuses only on different perspective on the origins of life. Scholars in the field of evolutionary religious studies have shed much light on the nature of humanity by exploring the evolutionary underpinnings of religion itself. And work in this area naturally connects with the field of positive psychology. Religions across time and place seek to provide guidance on living the good life—and the evolutionary perspective has provided great insights into the evolutionary function of religion. With this in mind, it is clear that work on the evolutionary functions of religion connects with the goals of positive psychology.

Research into the evolutionary origins of religion that explicitly takes the goals of positive psychology into account can help advance many important questions surrounding human nature. Sample research questions in this area are as follows: (a) Are there themes found that cut across the various religions that pertain to the pursuit of happiness? (b) What are the proximate goals of religious experiences when it comes to well-being? (c) What is the role of religion in cultivating healthy organizations?

The Evolutionary Psychology of Community

One of the goals of positive psychology is to shed light onto the factors that go into building healthy and effective human communities. Any individual exists in multiple communities at different levels at any given time, and taking steps to help increase trust, collaboration, and effectiveness within communities has the capacity to benefit people enormously. You might find yourself in the following communities: your neighborhood, your church, your local volunteer organization (e.g., Rotary Club), your department at work, a local political action group, etc. Given how communal we are as a species, it is important to understand the factors that lead to well-functioning communities.

The evolutionary perspective has much to say when it comes to the nature of community, including ideas on the origins of our tendency to organize and connect with others beyond lines of kinship. From this perspective, evolutionarily informed research on the topic of community might address such questions as (a) Are there factors that underlie successful community organization that cut across time and space? (b) Do religious communities differ from nonreligious communities in important ways? Can the evolutionary perspective on religion help inform this question? (c) Based on the work of Robin Dunbar and the idea of small-scale societies, do small-scale communities map better onto our natural psychology compared with large-scale communities?

Implications for Living

Life is hard—there is no denying this basic fact. The human emotion system is not perfect. People betray one another at times. We are regularly surrounded by supernormal, evolutionarily mismatched stimuli, such as baked goods, tobacco, and McDonald's, that entice us to make poor decisions that affect us adversely down the line. Human health is never perfect—and loss is an inherent part of living. Life is hard.

This all said, we are hopeful that this book provides some pathways to living that help make the ride positive and worth it. If you are reading this book as a self-help manual, this section is for you.

Physical Health

Work on human health that takes a Darwinian approach has been famously productive. A core theme in this area pertains to evolutionary mismatch—our bodies were not shaped by evolutionary forces to live in modern contexts. For the lion's share of human evolution, our ancestors ate only natural food (if food was available at all!), and these nomads exercised regularly out of necessity.

Implications: As much as you can, you should try to eat and exercise as our ancestors did. Some people make a big stink about the Paleo lifestyle and try to say that the idea is something of an overstatement. Perhaps this is so, but the facts are as follows: Eating only unprocessed food is a great way to stay in shape. Exercising by walking, running, and lifting weights—which mimics the daily routine of our ancestors—is also a great way to stay in shape. On the other hand, leading a sedentary life and eating a diet that is primarily comprised of processed foods is a recipe for obesity, cardiac failure, and Type 2 diabetes.

Mental Health

Our modern world has us surrounded with stimuli that mismatch our ancestral environments. And mental health problems famously track environments that are filled with instances of evolutionary mismatch.

To lead a life typified by mental hygiene, you'd be wise to make some life-style changes to help get your day-to-day environment relatively free of mismatch.

Implications: Humans evolved to be close to kin, so stay connected with your kin! From an evolutionary perspective, it is quite true that blood is thicker than is water. We evolved as a communal ape, so make sure that you play a role in your community: volunteer, join an organization, or-ganize events that bring people together, etc. From a Darwinian perspective, wealth is not based on money; rather, wealth is based on the connections that you have and the mark that you leave on the world. Finally, watch out for addictions. So many of the things that people are addicted to these days did not even exist under ancestral conditions—and we are addicted to them precisely because they reflect stimuli in high doses that would have been ap-pealing to our ancestors for evolutionary reasons. Take a break from your cell phone! And take steps to break any other addictions that you might have. Remember, tobacco, alcohol, and pornography are all postagrarian techno-logical advances that did not exist during the lion's share of human evolution. Treat with some level of skepticism any technology that is evolutionarily unnatural.

Relationships

Humans evolved to have several classes of strong relationships with others. We evolved to have special relationships with kin, but we also evolved as a partly coalitional ape, forming important alliances with people outside one's kin group. We also evolved a distinctive set of adaptations related to mating and intimate relationships, and cultivating positive relationships on this front is critical for effective functioning.

Implications: Getting along with others is an essential part of being human. Seeing past flaws of others in many cases is critical in forming im-portant relationships that are foundational to success in all facets of life. When it comes to intimate relationships, realize that mating systems that approximate monogamy run relatively deeply in our species. While mo-nogamy is not the only game in town when it comes to human mating, it is very common and makes a good bit of evolutionary sense vis-à-vis the importance of biparental care connected with childrearing. So don't take your partner for granted—and always try to follow the Golden Rule when it comes to those who are closest to you in your world. There will be long-term benefits to such an approach.

Parenting

From an evolutionary perspective, parenting is as critical a domain as any when it comes to all aspects of living. Our offspring are our ultimate vehicles for getting our genes into the next generation. Kids are evolutionary product sine qua non when it comes to being human.

Implications: Don't take parenting for granted. And in raising your kids, always keep an eye on the issue of other-orientedness. Humans evolved to be other-oriented, but this is largely mediated through social learning and good parenting. Kids need to learn that they are part of something bigger and that other-oriented behaviors that have short-term costs are likely to have long-term benefits. And make sure to have fun with your kids because they do grow up in the blink of an eye.

Community

From the work of Bingham and Souza (2009) and Wilson (2007), we know that humans are communal apes. Forming communities that often cut across kin lines is a foundational aspect of being human. There is nothing more rewarding than playing an important role in a well-functioning community. Working together, a group of individuals can achieve nearly anything. And there are so many different communities out there that getting involved in some community in an important way is easy to do. Get involved, volunteer, and take part in something that reminds you of the fact that you are part of something greater. You will benefit as a result—and you will help others along the way.

Implications: Humans are communal apes, so it is important to make sure that you are connected to others in communal ways. If you are having trouble fitting in with some community, it would likely benefit you to be proactive, looking for organizations in your region, or even online, that relate to your interests and experiences. There are tons of organizations and communities out there, and they are nearly all looking to increase membership. Remember that you are a communal ape, so take advantage of this fact!

Social Connections

Humans evolved to have a strong foundation of social connections. Under ancestral conditions, these connections existed within small, tight-knit communities. Our minds evolved to exist in such a context.

Implications: Be careful when it comes to cutting people out of your life. While some acts may be truly unforgivable, social estrangements were very costly for our ancestors during human evolution, and modern responses to estrangements clearly show the vestiges of this ancestral reality. Forgiveness is a critical feature of our evolved psychology. And at the end of the day, let's face it—none of us is perfect. None of us is even close. People who have fewer social estrangements in their worlds do better in so many ways than do those who have many estrangements. There is a lesson in there.

Religion and Spirituality

Religion is not perfect, but it is a basic feature of being human. Religion likely evolved to keep behaviors within a community in check to cultivate behavioral patterns in which people would work toward the common good. There are, of course, many kinds of human communities, such as Humanist groups, that focus on an other-oriented approach without the inclusion of formal religion. This said, it seems that humans need community—whether it comes from religion or otherwise.

Implications: Social connections and community are critical features of living the good life from an evolutionary perspective. Religions likely emerged partly as mechanisms to provide community and to cultivate an other-oriented approach to life. If you are religious, understanding this important function of religion may well help you get the most out of your religious experiences. And if you're not religious, as is likely true of many readers of this book, note that developing important social connections with others and cultivating an other-oriented approach to life is, based on our evolutionary heritage, critical to getting the most out of life.

Bottom Line

Positive psychology is concerned with best understanding how people can live the good life—by advancing our understanding of personal and community well-being. Based on Darwin's big ideas, evolutionary psychology is focused on using evolutionary principles to help us best understand all aspects of human behavior. Positive evolutionary psychology, introduced here, is our effort to marry these two fields—providing a framework for thinking about how the goals of positive psychology can be elucidated by the field of evolutionary psychology.

At the end of the day, we all seek answers. Life is fleeting, and we are all motivated to make it count. Positive psychology focuses largely on how we can cultivate the positive aspects of the human experience. Evolutionary psychology is a remarkably powerful set of intellectual tools designed to help provide insights into the entirety of human behavior. Positive evolutionary psychology, presented here, is the marriage of these two fields—designed with the ultimate goal of using Darwin's powerful insights to help improve the human condition.

Thank you for taking the time to read this book, and here is to our shared future.

References

Bingham, P. M., & Souza, J. (2009). *Death from a distance and the birth of a humane universe*. Lexington, KY: BookSurge.

Mayr, E., & Provine, W. (Eds.). (1980). *The evolutionary synthesis*. Cambridge, MA: Harvard University Press.

Wilson, D. S. (2002). *Darwin's cathedral: Evolution, religion and the nature of society*. Chicago: University of Chicago Press.

Wilson, D. S. (2007). *Evolution for everyone: How Darwin's theory can change the way we think about our lives*. New York: Delacorte Press.

Index

Note: Tables are indicated by *t* following the page number.

For the benefit of digital users, indexed terms that span two pages (e.g., 52–53) may, on occasion, appear on only one of those pages.